Research Notes in Mathematics

Submission of proposals for consideration
Suggestions for publication, in the form of outlines and representative samples, are invited by the editorial board for assessment. Intending authors should contact either the main editor or another member of the editorial board, citing the relevant AMS subject classifications. Refereeing is by members of the board and other mathematical authorities in the topic concerned, located throughout the world.

Preparation of accepted manuscripts
On acceptance of a proposal, the publisher will supply full instructions for the preparation of manuscripts in a form suitable for direct photo-lithographic reproduction. Specially printed grid sheets are provided and a contribution is offered by the publisher towards the cost of typing.

Illustrations should be prepared by the authors, ready for direct reproduction without further improvement. The use of hand-drawn symbols should be avoided wherever possible, in order to maintain maximum clarity of the text.

The publisher will be pleased to give any guidance necessary during the preparation of a typescript, and will be happy to answer any queries.

Important note
In order to avoid later retyping, intending authors are strongly urged not to begin final preparation of a typescript before receiving the publisher's guidelines and special paper. In this way it is hoped to preserve the uniform appearance of the series.

Titles in this series

1 Improperly posed boundary value problems
A Carasso and A P Stone

2 Lie algebras generated by finite dimensional ideals
I N Stewart

3 Bifurcation problems in nonlinear elasticity
R W Dickey

4 Partial differential equations in the complex domain
D L Colton

5 Quasilinear hyperbolic systems and waves
A Jeffrey

6 Solution of boundary value problems by the method of integral operators
D L Colton

7 Taylor expansions and catastrophes
T Poston and I N Stewart

8 Function theoretic methods in differential equations
R P Gilbert and R J Weinacht

9 Differential topology with a view to applications
D R J Chillingworth

10 Characteristic classes of foliations
H V Pittie

11 Stochastic integration and generalized martingales
A U Kussmaul

12 Zeta-functions: An introduction to algebraic geometry
A D Thomas

13 Explicit *a priori* inequalities with applications to boundary value problems
V G Sigillito

14 Nonlinear diffusion
W E Fitzgibbon III and H F Walker

15 Unsolved problems concerning lattice points
J Hammer

16 Edge-colourings of graphs
S Fiorini and R J Wilson

17 Nonlinear analysis and mechanics: Heriot-Watt Symposium Volume I
R J Knops

18 Actions of fine abelian groups
C Kosniowski

19 Closed graph theorems and webbed spaces
M De Wilde

20 Singular perturbation techniques applied to integro-differential equations
H Grabmüller

21 Retarded functional differential equations: A global point of view
S E A Mohammed

22 Multiparameter spectral theory in Hilbert space
B D Sleeman

24 Mathematical modelling techniques
R Aris

25 Singular points of smooth mappings
C G Gibson

26 Nonlinear evolution equations solvable by the spectral transform
F Calogero

27 Nonlinear analysis and mechanics: Heriot-Watt Symposium Volume II
R J Knops

28 Constructive functional analysis
D S Bridges

29 Elongational flows: Aspects of the behaviour of model elasticoviscous fluids
C J S Petrie

30 Nonlinear analysis and mechanics: Heriot-Watt Symposium Volume III
R J Knops

31 Fractional calculus and integral transforms of generalized functions
A C McBride

32 Complex manifold techniques in theoretical physics
D E Lerner and P D Sommers

33 Hilbert's third problem: scissors congruence
C-H Sah

34 Graph theory and combinatorics
R J Wilson

35 The Tricomi equation with applications to the theory of plane transonic flow
A R Manwell

36 Abstract differential equations
S D Zaidman

37 Advances in twistor theory
L P Hughston and R S Ward

38 Operator theory and functional analysis
I Erdelyi

39 Nonlinear analysis and mechanics: Heriot-Watt Symposium Volume IV
R J Knops

40 Singular systems of differential equations
S L Campbell

41 N-dimensional crystallography
R L E Schwarzenberger

42 Nonlinear partial differential equations in physical problems
D Graffi

43 Shifts and periodicity for right invertible operators
D Przeworska-Rolewicz

44 Rings with chain conditions
A W Chatters and C R Hajarnavis

45 Moduli, deformations and classifications of compact complex manifolds
D Sundararaman

46 Nonlinear problems of analysis in geometry and mechanics
M Atteia, D Bancel and I Gumowski

47 Algorithmic methods in optimal control
W A Gruver and E Sachs

48 Abstract Cauchy problems and functional differential equations
F Kappel and W Schappacher

49 Sequence spaces
W H Ruckle

50 Recent contributions to nonlinear partial differential equations
H Berestycki and H Brezis

51 Subnormal operators
J B Conway

52 Wave propagation in viscoelastic media
F Mainardi

53 Nonlinear partial differential equations and their applications: Collège de France Seminar. Volume I
H Brezis and J L Lions

54 Geometry of Coxeter groups
H Hiller

55 Cusps of Gauss mappings
T Banchoff, T Gaffney and C McCrory

56 An approach to algebraic K-theory
A J Berrick
57 Convex analysis and optimization
J-P Aubin and R B Vintner
58 Convex analysis with applications in the differentiation of convex functions
J R Giles
59 Weak and variational methods for moving boundary problems
C M Elliott and J R Ockendon
60 Nonlinear partial differential equations and their applications: Collège de France Seminar. Volume II
H Brezis and J L Lions
61 Singular systems of differential equations II
S L Campbell
62 Rates of convergence in the central limit theorem
Peter Hall
63 Solution of differential equations by means of one-parameter groups
J M Hill
64 Hankel operators on Hilbert space
S C Power
65 Schrödinger-type operators with continuous spectra
M S P Eastham and H Kalf
66 Recent applications of generalized inverses
S L Campbell
67 Riesz and Fredholm theory in Banach algebra
B A Barnes, G J Murphy, M R F Smyth and T T West
68 Evolution equations and their applications
F Kappel and W Schappacher
69 Generalized solutions of Hamilton-Jacobi equations
P L Lions
70 Nonlinear partial differential equations and their applications: Collège de France Seminar. Volume III
H Brezis and J L Lions
71 Spectral theory and wave operators for the Schrödinger equation
A M Berthier
72 Approximation of Hilbert space operators I
D A Herrero
73 Vector valued Nevanlinna Theory
H J W Ziegler
74 Instability, nonexistence and weighted energy methods in fluid dynamics and related theories
B Straughan
75 Local bifurcation and symmetry
A Vanderbauwhede
76 Clifford analysis
F Brackx, R Delanghe and F Sommen
77 Nonlinear equivalence, reduction of PDEs to ODEs and fast convergent numerical methods
E E Rosinger
78 Free boundary problems, theory and applications. Volume I
A Fasano and M Primicerio
79 Free boundary problems, theory and applications. Volume II
A Fasano and M Primicerio
80 Symplectic geometry
A Crumeyrolle and J Grifone
81 An algorithmic analysis of a communication model with retransmission of flawed messages
D M Lucantoni
82 Geometric games and their applications
W H Ruckle
83 Additive groups of rings
S Feigelstock
84 Nonlinear partial differential equations and their applications: Collège de France Seminar. Volume IV
H Brezis and J L Lions
85 Multiplicative functionals on topological algebras
T Husain
86 Hamilton-Jacobi equations in Hilbert spaces
V Barbu and G Da Prato
87 Harmonic maps with symmetry, harmonic morphisms and deformations of metrics
P Baird
88 Similarity solutions of nonlinear partial differential equations
L Dresner
89 Contributions to nonlinear partial differential equations
C Bardos, A Damlamian, J I Díaz and J Hernández
90 Banach and Hilbert spaces of vector-valued functions
J Burbea and P Masani
91 Control and observation of neutral systems
D Salamon
92 Banach bundles, Banach modules and automorphisms of C*-algebras
M J Dupré and R M Gillette
93 Nonlinear partial differential equations and their applications: Collège de France Seminar. Volume V
H Brezis and J L Lions
94 Computer algebra in applied mathematics: an introduction to MACSYMA
R H Rand
95 Advances in nonlinear waves. Volume I
L Debnath
96 FC-groups
M J Tomkinson
97 Topics in relaxation and ellipsoidal methods
M Akgül
98 Analogue of the group algebra for topological semigroups
H Dzinotyiweyi
99 Stochastic functional differential equations
S E A Mohammed
100 Optimal control of variational inequalities
V Barbu
101 Partial differential equations and dynamical systems
W E Fitzgibbon III
102 Approximation of Hilbert space operators. Volume II
C Apostol, L A Fialkow, D A Herrero and D Voiculescu
103 Nondiscrete induction and iterative processes
V Ptak and F-A Potra
104 Analytic functions – growth aspects
O P Juneja and G P Kapoor
105 Theory of Tikhonov regularization for Fredholm equations of the first kind
C W Groetsch

106 Nonlinear partial differential equations and free boundaries
J I Díaz

107 Tight and taut immersions of manifolds
T E Cecil and P J Ryan

108 A layering method for viscous, incompressible L_p flows occupying R^n
A Douglis and E B Fabes

109 Nonlinear partial differential equations and their applications: Collège de France Seminar. Volume VI
H Brezis and J L Lions

110 Finite generalized quadrangles
S E Payne and J A Thas

111 Advances in nonlinear waves. Volume II
L Debnath

112 Topics in several complex variables
E Ramírez de Arellano and D Sundararaman

113 Differential equations, flow invariance and applications
N H Pavel

114 Geometrical combinatorics
F C Holroyd and R J Wilson

115 Generators of strongly continuous semigroups
J A van Casteren

116 Growth of algebras and Gelfand–Kirillov dimension
G R Krause and T H Lenagan

117 Theory of bases and cones
P K Kamthan and M Gupta

118 Linear groups and permutations
A R Camina and E A Whelan

119 General Wiener–Hopf factorization methods
F-O Speck

120 Free boundary problems: applications and theory, Volume III
A Bossavit, A Damlamian and M Fremond

121 Free boundary problems: applications and theory, Volume IV
A Bossavit, A Damlamian and M Fremond

122 Nonlinear partial differential equations and their applications: Collège de France Seminar. Volume VII
H Brezis and J L Lions

123 Geometric methods in operator algebras
H Araki and E G Effros

124 Infinite dimensional analysis – stochastic processes
S Albeverio

Differential equations, flow invariance and applications

N H Pavel

University of Iasi

Differential equations, flow invariance and applications

Pitman Advanced Publishing Program

BOSTON · LONDON · MELBOURNE

PITMAN PUBLISHING LIMITED
128 Long Acre, London WC2E 9AN

PITMAN PUBLISHING INC
1020 Plain Street, Marshfield, Massachusetts 02050

Associated Companies
Pitman Publishing Pty Ltd, Melbourne
Pitman Publishing New Zealand Ltd, Wellington
Copp Clark Pitman, Toronto

First published 1984

AMS Subject Classifications: (main) 34G20, 47H15, 47H05–06
(subsidiary) 47F05, 35R20, 35A35

ISSN 0743-0337

Library of Congress Cataloging in Publication Data

Pavel, N. H. (Nicolae H.)
Differential equations, flow invariance and applications.

(Research notes in mathematics; 113)
"Pitman advanced publishing program."
Bibliography: p.
Includes index.
1. Differential equations, Nonlinear. 2. Operator equations, Nonlinear. 3. Nonlinear operators. I. Title. II. Series.
QA372.P36 1984 515.3′55 84-14811
ISBN 0-273-08651-0

British Library Cataloguing in Publication Data

Pavel, N. H.
Differential equations, flow invariance and applications.—(Research notes in mathematics, ISSN 0743–0337; 113)
1. Differential equations
I. Title II. Series
515.3′5 QA371

ISBN 0-273-08651-0

Reproduced and printed by photolithography
in Great Britain by Biddles Ltd, Guildford

Contents

Preface

The purpose of this work is to present in a coherent way some of the main recent research on differential equations on closed (or merely locally closed) subsets, and then to give various applications.

After an introduction to the preliminaries of nonlinear analysis in Chapter 1, Chapter 2 deals with some fundamental results on differential equations associated with continuous and dissipative time-dependent domain operators. The special case of Lipschitz condition is treated in a simpler manner.

Chapter 3 (as well as some sections of Chapter 2) is devoted to some applications of the theory of the flow-invariance of a set with respect to a differential equation. Chapter 4 shows that a Banach manifold is the most natural framework in which to treat flow-invariance problems. By using the semigroup approach, Chapter 5 presents some general results on perturbed differential equations, extending those in Chapter 2. Some applications to partial differential equations are also given.

Classical existence and uniqueness results on the Cauchy problem for ordinary differential equations are included. Moreover, since the interior of our closed subsets here may be empty, the results can also be applied to partial differential equations. The very general notion of tangency that we use in the book is also important in optimization. This aspect is pointed out in Chapter 4.

Most of the material is self-contained and the only prerequisite is an elementary course in functional analysis. The book is based on the second volume of the lecture notes that I wrote in 1980-82, at the University of Iasi, which were intended only for my students in the Mathematics Department. In the present work, much has been added, or improved and expanded: there are some new sections in Chapter 2, Chapter 4 is a new chapter, Chapter 5 is completely rewritten, and so on.

My interest in this very important topic goes back to some conversations, in 1973, with Professors Brezis, Browder and Crandall. Some sections of Chapter 5 have been improved following suggestions by Professors Barbu and Pazy. My collaboration with Professors Ursescu, Morosanu, Motreanu and

Vrabie has contributed greatly to the improvement of many results given in the book. I am grateful to all of these friends and advisers.

I am also indebted to Pitman Publishing for undertaking the publication of this book, and for very pleasant cooperation in its preparation.

N H Pavel

Iasi
April, 1984

Notation and general remarks

The symbols that are used in this book are defined where they occur for the first time. Here we point out only a few of them.

X denotes a Banach space over the field of real numbers R. X^* is the dual of X. The value of $x^* \in X^*$ at $x \in X$ is denoted by $\langle x,x^*\rangle$, or $\langle x^*,x\rangle$ (or even $x^*(x)$).

If $a,b \in R$, $a < b$, then $[a,b]$ and $]a,b[$ denote the closed and the open interval, respectively.

$x \equiv y$ means: "x equals y by definition": if $x,y \in X$ then $[x,y] \equiv \{\lambda x + (1-\lambda)y;\ \lambda \in [0,1]\}$.

If X is a complex Banach space, then in Proposition 3.2, Ch. 1, we have to replace $\langle w,y_1-y_2\rangle \leq 0$ by $\mathrm{Re}\ \langle w,y_1-y_2\rangle \leq 0$, where Re z is the real part of the complex number z.

For Chapter 4, $D(g)_x{}^{-1}(0) \equiv (D(g)_x)^{-1}(0)$.

Theorems (propositions, corollaries etc.) and displayed formulae are numbered as follows: Theorem 2.1 is the first theorem in Section 2 of a chapter (say Chapter 2). References to this theorem within Chapter 2 simply read "Theorem 2.1". Where it is referred to in other chapters, it is cited as "Theorem 2.1, Ch. 2", and so on.

1 Preliminaries of nonlinear analysis

The aim of this chapter is to introduce the reader to the basic properties of nonlinear (possible multivalued) operators of dissipative and accretive type (Section 3). These operators are important in the theory of ordinary differential equations and partial differential equations. They are more general than Lipschitz operators.

Some properties of duality mapping are presented in Section 2. The results of Section 2 (especially Lemma 2.1) are also necessary in the theory of differential equations associated with dissipative operators (Chapter 2).

1. Characterization of strictly convex spaces. Convergence and projection in uniformly convex spaces.

Definition 1.1 (1) *A normed space* X *is said to be strictly convex if, for each* $x,y \in X$ *with* $\|x\| = \|y\| = r$, *and* $0 < \lambda < 1$, $r > 0$, *one has* $\|\lambda x + (1-\lambda)y\| < r$.
(2) X *is said to be uniformly convex if, for each* $\varepsilon \in]0,2]$, *there is* $\delta = \delta(\varepsilon) > 0$ *such that, whenever* $x, y \in X$ *with* $\|x\| = \|y\| = 1$ *and* $\|x-y\| > \varepsilon$, *it follows that* $\|x + y\| /2 \leqslant 1-\delta$.

Obviously, a uniformly convex space is strictly convex. The converse implication is not valid. It is well known that any uniformly convex space is reflexive and consequently it is a Banach space.

In connection with strictly convex spaces we shall present only those properties that are needed in this book. Set

$$S(x,r) = \{y \in X, \ \|y - x\| = r\}$$
$$B(x,r) = \{y \in X, \ \|y - x\| \leqslant r\}. \tag{1.1}$$

Definition 1.2 *The spheres* $S(x_i,r_i) \subset X$, $i = 1,2$, *are said to be exterior tangent if* $\|x_1-x_2\| = r_1 + r_2$.

The intersection of two exterior tangent spheres $S(x_i,r_i)$, $i = 1,2$, is always nonempty. Indeed, for $\lambda_0 = r_2/(r_1+r_2)$, it follows that

$$x_o \equiv \lambda_o x_1 + (1-\lambda_o)x_2 \in S(x_1,r_1) \cap S(x_2,r_2). \tag{1.1)'$$

In general, the intersection of two exterior tangent spheres may contain more than one point. For example, let us consider $x_1^o = (1,0)$ and $x_2^o = (-1,0)$ in R^2. If R^2 is endowed with the norm

$$\|(a,b)\| = \max\{|a|, |b|\}, \ (a,b) \in R^2$$

then the intersection of the spheres $S(x_i^o,1)$, $i = 1,2$, contains the whole segment $[(0,-1), (0,1)] \subset R^2$. Therefore, the following result is significant.

Proposition 1.1 *A normed space* X *is strictly convex iff the intersection of any two exterior tangent spheres consists of a single point.*

Proof Let X be strictly convex and let $S(x_i,r_i)$, $i = 1,2$, be two exterior tangent spheres. Then it follows that

$$S(x_1,r_1) \cap S(x_2,r_2) = \{x_o\} \tag{1.2}$$

where x_o is given by (1.1)'. Indeed, if we assume that there is one more point $x \neq x_o$ with $x \in S(x_1,r_1) \cap S(x_2,r_2)$, then we have

$$\|(x_o + x)/2 - x_i\| < r_i, \quad i = 1,2.$$

This implies

$$r_1 + r_2 \leq \|x_1-(x_o + x)/2\| + \|x_2-(x_o + x)/2\| < r_1 + r_2$$

which is a contradiction. Conversely, if we assume that for two arbitrary $S(x_i,r_i)$, (1.2) holds, then X is necessarily strictly convex. Indeed, if this is not the case, then there is $S(x_1,r_1)$ which contains a line segment $[y_1,y_2]$ with $y_i \in S(x_1,r_1)$, $i = 1,2$, and $y_1 \neq y_2$. Set $y_o = (y_1 + y_2)/2$ and $x_2 = 2y_o-x_1$. It is easy to verify that the spheres $S(x_i,r_1)$, $i = 1,2$, are exterior tangent and their intersection contains $[y_1,y_2]$ (which contradicts hypothesis (1.2)).

On uniformly convex spaces the following useful property holds.

Lemma 1.1 *Let* X *be a uniformly convex space and* $\{x_n\} \subset X$. *If* $x_n \rightharpoonup x$ *as* $n \to \infty$ *and* $\limsup_{n \to \infty} \|x_n\| \leq \|x\|$ *(i.e.* $\|x_n\| \to \|x\|$*), then* $x_n \to x$.

Proof If $x = 0$ the assertion is trivial, hence let us assume that $x \neq 0$ and that $x_n \neq 0$ for all $n = 1,2,\ldots$. Set $y_n = x_n/\|x_n\|$, $y = x/\|x\|$ and $a_n = 1 - \|x\|/\|x_n\|$. Since $a_n \to 0$ and

$$y_n - y = (1-a_n)x_n\|x\|^{-1} - x\|x\|^{-1}$$

we see that $y_n \rightharpoonup y$ as $n \to \infty$. Assume that $x_n \not\to x$. Then there is $\varepsilon > 0$ such that $\|y_{n'}-y\| \geq \varepsilon$ for infinitely many natural numbers n'. By the uniform convexity of X, there is $\delta > 0$ such that $\|y_{n'} + y\| \leq 2(1-\delta)$. Take $y^* \in X^*$ with the properties: $y^*(y) = 1 = \|y^*\|$. Then $y^*(y_{n'} + y) \to 2y^*(y) = 2$. On the other hand, $|y^*(y_{n'} + y)| \leq \|y_{n'} + y)\| \leq 2(1-\delta)$, which yields the contradiction $2 \leq 2(1-\delta)$. □

Using Lemma 1.1 we can easily prove the next result.

Proposition 1.2 *Let us consider the sequences* $\{a_n^i\} \subset R$, $i = 1,2$, *strictly decreasing to zero as* $n \to \infty$. *If* X *is uniformly convex and* $\{\tilde{x}_n\} \subset X$ *has the property that*

$$\tilde{x}_n \in B(x_1,r_1 + a_n^1) \cap B(x_2,r_2 + a_n^2) \equiv C_n$$

for all $n = 1,2 \ldots$ *and* $\|x_1-x_2\| = r_1 + r_2$, *then*

$$\lim_{n\to\infty} \tilde{x}_n = \frac{r_2}{r_1 + r_2}\, x_1 + \frac{r_1}{r_1 + r_2}\, x_2 \equiv x_0.$$

Proof Since a_n^i is strictly decreasing to zero, it is clear that $C_n \supset C_{n+1}$, $n = 1,2,\ldots$. Therefore $\tilde{x}_{n+p} \in C_{n+p} \subset C_n$ for all $p = 1,2,\ldots$. On the other hand, C_n is a bounded closed convex subset of X, and X is reflexive. Thus we may assume that $\{\tilde{x}_{n+p}\}_{p=1}^{\infty}$ (which is bounded) is weakly convergent to an element x. Since C_n is weakly closed, it follows that $x \in C_n$. In fact one has $\tilde{x}_n \rightharpoonup x \in \bigcap_{n\in N} C_n$. Since $S(x_i,r_i)$, $i = 1,2$, are assumed to be exterior tangent, it follows from Proposition 1.1 that $\bigcap_{n\in N} C_n = \{x_0\}$. Thus we have $x = x_0$, hence $\tilde{x}_n \rightharpoonup x_0$ as $n \to \infty$. But $\tilde{x}_n \in B(x_1,r_1 + a_n^1)$, which means $\|\tilde{x}_n - x_1\| \leq r_1 + a_n^1$ and therefore

$$\limsup_{n \to \infty} \|\tilde{x}_n - x_1\| \leq r_1 = \|x_0 - x_1\|.$$

Combining this inequality, the property $\tilde{x}_n - x_1 \rightharpoonup x_0 - x_1$ and Lemma 1.1, we see that $\tilde{x}_n - x_1 \to x_0 - x_1$ as $n \to \infty$.

Proposition 1.3 *Let* A *be a closed convex subset of the uniformly convex space* X. *Then the following properties hold:*

(a) *for each* $y \in X$, *there is a unique element (call it* Py*) of* A *such that* $d(y;A) = \|y - Py\|$ ($d(y;A)$ = the distance from y to A);

(b) *the projection map* P *of* X *onto* A *is continuous.*

Proof (a) Fix $y \in X$. There is $\{x_n\} \subset A$ such that $\lim_n \|y - x_n\| = d(y;A) \equiv d$. We now prove that $\{x_n\}$ is a Cauchy sequence. The convexity of A implies $\frac{1}{2}(x_n + x_m) \in A$ for all n and m, therefore

$$\|\tfrac{1}{2}(x_n - y) + \tfrac{1}{2}(x_m - y)\| = \|\tfrac{1}{2}(x_n + x_m) - y\| \geq d.$$

Thus we have

$$\lim_n \|x_n - y\| = \lim_m \|x_m - y\| = d, \quad \lim_{m,n} \|(x_n - y) + (x_m - y)\| = 2d.$$

Since X is uniformly convex, it follows that $\lim_{n,m} \|(x_n - y) - (x_m - y)\| = 0$, hence $\{x_n\}$ is strongly convergent. Set $Py = \lim_n x_n$. Then $Py \in A$ (since A is closed). The uniqueness of Py is a consequence of the convexity of the subset

$$\{z \in A, \|y - z\| = d\} \subset S(y,d)$$

and of strict convexity of X.

(b) Let $\{x_n\} \subset X$ with $x_n \to x$. We have to prove that $Px_n \to Px$. Recall that $|d(x_n;A) - d(x;A)| \leq \|x_n - x\|$, that is, $\big|\|x_n - Px_n\| - \|x - Px\|\big| \leq \|x_n - x\|$. Hence $\lim_n \|x_n - Px_n\| = \|x - Px\|$, which implies the boundedness of Px_n. We may assume that Px_n is weakly convergent to an element $u \in A$. Thus $x_n - Px_n \rightharpoonup x - u$. On the other hand, $\|x - u\| \leq \liminf_{n\to\infty} \|x_n - Px_n\| =$

$\|x - Px\| = d(x;A)$, which gives $u = Px$. Finally, by Lemma 1.1 it follows that $x_n - Px_n \to x - Px$, which implies $Px_n \to Px$. □

Remark 1.1 Assertion (a) of Proposition 1.3 remains valid if X is only reflexive and strictly convex. Indeed, since $\|x_n - y\| \to d$, it follows that $\{x_n\}$ is bounded. Since X is reflexive we may assume (relabelling if necessary) that $x_n \rightharpoonup x$. But A is weakly closed, so $x \in A$. On the other hand, $y - x_n \rightharpoonup y - x$ implies

$$\|y - x\| \leq \lim_n \|y - x_n\| = d(y,A)$$

therefore $x = Py$. □

In view of the above remarks, we have the following.

Corollary 1.1 *Assume that* X *is reflexive and* $A \subset X$ *is a closed and convex subset of* X. *Then* A *has an element of minimal norm. If in addition* X *is strictly convex, then* P0 *is the (unique) element of* A, *of minimal norm, i.e.* $\|P0\| \leq \|x\|$, *for all* $x \in A$ $(P0 \in A)$.

§2. Duality mapping

2.1 Definitions and basic properties. The directional derivative of the norm

Let X be a normed space over R. Denote by $\|\cdot\|$ both the norm of X and the norm of X*. The function F from X into 2^{X^*} defined by

$$F(x) = \{x^* \in X^*;\ x^*(x) = \|x\|^2 = \|x^*\|^2\} \tag{2.1}$$

is called the *duality mapping* of X. The fact that F(x) is nonempty is a consequence of the Hahn-Banach theorem.

Definition 2.1 *Let* Y *be a normed space and let:* $D(A) \subset X \to 2^Y$ *be a multi-valued mapping.*

(1) A *is said to be demiclosed if for every* $\{x_n\} \subset D(A)$ *and* $\{y_n\} \subset Y$ *with* $y_n \in Ax_n$, $x_n \to x$ *and* $y_n \rightharpoonup y$, *it follows* $x \in D(A)$ *and* $y \in Ax$.

(2) $F: X \to X^*$ *is said to be demicontinuous if* F *is continuous from the strong topology of* X *to the weak* topology of* X*.

__Proposition 2.1__ (i) *If* X *is an arbitrary normed space, then its duality mapping* F *is demiclosed.*

(ii) *If* X *is strictly convex then* F *is single-valued and demicontinuous.*

__Proof__ (i) Let $\{x_n\} \subset X$ and $\{y_n^*\} \subset X^*$ with the properties $y_n^* \in F(x_n)$ (hence $y_n^*(x_n) = \|x_n\|^2 = \|y_n^*\|^2$), $x_n \to x$ while $y_n^* \rightharpoonup y^*$. First, it follows that $y_n^*(x_n) \to y^*(x)$ as $n \to \infty$, which implies $y^*(x) = \|x\|^2$. We always have

$$\|y^*\| \leqslant \liminf_{n \to \infty} \|y_n^*\|. \tag{2.2}$$

We now see that (2.2) gives $\|y^*\| \leqslant \|x\|$, while $y^*(x) = \|x\|^2$ gives the converse inequality (and therefore $y^* \in F(x)$).

(ii) It is clear that for each $x \in X$, $F(x)$ is a convex subset of X^* and $F(x) \subset S^*(\|x\|)$, where

$$S^*(r) = \{x^* \in X^*, \|x^*\| = r\}, \; r > 0. \tag{2.3}$$

If we assume that X^* is strictly convex, then $F(x)$ consists of a single element $x^* \in X^*$.

Furthermore, let $\{x_n\} \subset X$ be strongly convergent to x. Since $\{x_n\}$ and x are contained in a separable subspace of X, we may assume (without loss of generality) that even X is separable. In this case, the weak* topology of X relative to a bounded subset is metrizable. Since the unit ball of X is weakly* compact we may assume that $F(x_n)$ is weakly convergent to $y^* \in X^*$. Arguing as in (i), we prove that $y^* = F(x)$, therefore every weak star limit point of $F(x_n)$ coincides with $F(x)$ (i.e. $F(x_n) \longrightarrow F(x)$ weakly-star).

__Proposition 2.2__ *Assume that* X *is a real Banach space and* X* *is uniformly convex. Then the duality mapping* $F: X \to X^*$ *is uniformly continuous on bounded subsets of* X.

__Proof__ Suppose, by contradiction, that there are $\{x_n\}$ and $\{y_n\} \subset X$ with the properties:

$$\|x_n\| \leqslant a, \; \|y_n\| \leqslant a, \; \|x_n - y_n\| \to 0, \; \|F(x_n) - F(y_n)\| \geqslant \varepsilon > 0 \tag{2.4}$$

for all n = 1,2,... where a is a positive number. The last inequality of

(2.4) and $\|F(x_n)\| = \|x_n\|$, $\|F(y_n)\| = \|y_n\|$ show that neither x_n nor y_n are convergent to zero. Therefore we may assume that $\|x_n\| \geq b > 0$, $\|y_n\| \geq b$ for all n.

Set $u_n = x_n/\|x_n\|$, $v_n = y_n/\|y_n\|$. We have

$$\|u_n - v_n\| = \frac{\|(x_n-y_n)\|y_n\| + y_n(\|y_n\| - \|x_n\|)\|}{\|x_n\|\ \|y_n\|} \leq \frac{2\|y_n\|\ \|x_n-y_n\|}{b^2}$$

therefore

$$\lim_{n\to\infty} \|u_n - v_n\| = 0. \tag{2.5}$$

Obviously,

$$(F(u_n) + F(v_n))(u_n) = \|u_n\|^2 + \|v_n\|^2 + F(v_n)(u_n - v_n) \geq 2 - \|u_n - v_n\|$$

which implies

$$\frac{1}{2}\|F(u_n) + F(v_n)\| \geq 1 - \frac{1}{2}\|u_n - v_n\|. \tag{2.6}$$

Combining $\|F(u_n)\| = \|u_n\| = 1$, $\|F(v_n)\| = 1$, (2.5), (2.6) and the uniform convexity of X*, it follows at once that

$$\lim_{n\to\infty} (F(u_n) - F(v_n)) = 0 \text{ in } X^* \text{ (strongly)}.$$

On the other hand,

$$F(x_n) - F(y_n) = \|x_n\|(F(u_n) - F(v_n)) + (\|x_n\| - \|y_n\|)F(v_n) \tag{2.7}$$

where the property

$$F(\lambda x) = \lambda F(x),\ \lambda > 0 \tag{2.8}$$

has been used. The fact that F is positively homogeneous (i.e. (2.8)) follows directly from definition.

From (2.7) we get $\lim_{n\to\infty} \|F(x_n) - F(y_n)\| = 0$ which (in view of (2.4)) is a contradiction.

For the presentation of the connection between the duality mapping and the subdifferential of a convex function, some preliminaries are needed.

Let us define the following function (from $X \times X$ into R):

$$\langle y,x\rangle_s = \lim_{t\downarrow 0} \frac{\|x+ty\|^2 - \|x\|^2}{2t} \; ; \; \langle y,x\rangle_+ = \lim_{t\downarrow 0} \frac{\|x+ty\| - \|x\|}{t} \quad (2.9)$$

$$\langle y,x\rangle_i = \lim_{t\uparrow 0} \frac{\|x+ty\|^2 - \|x\|^2}{2t} \; ; \; \langle y,x\rangle_- = \lim_{t\uparrow 0} \frac{\|x+ty\| - \|x\|}{t} \quad (2.10)$$

where $x,y \in X$ and $t \in R$, $t \neq 0$.

These functions are well-defined since both $t \to \|x + ty\|^2$ and $t \to \|x + ty\|$ are convex functions.

Also, set

$$\langle y,x\rangle_t = \frac{\|x+ty\| - \|x\|}{t}, \; t \neq 0, \; x,y \in X. \quad (2.11)$$

Obviously

$$\left.\begin{array}{l} \langle y,x\rangle_s = \|x\| \langle y,x\rangle_+, \; \langle y,x\rangle_i = \|x\| \langle y,x\rangle_- \\ \langle y,x\rangle_+ \leqslant \langle y,x\rangle_t \leqslant \|y\|, \; t > 0 \\ \langle y,x\rangle_t \leqslant \langle y,x\rangle_-, \; t < 0 \end{array}\right\} \quad (2.12)$$

for all $x,y \in X$.

<u>Proposition 2.3</u> *For each* x *and* $y \in X$ *there are* $w_i \in F(x)$, $i = 1,2$ *such that*

$$\langle y,x\rangle_s = w_1(y), \; \langle y,x\rangle_i = w_2(y). \quad (2.13)$$

<u>Proof</u> Let us consider the linear subspace E of X spanned by x and y, that is

$$E = \{z \in X; \; z = ax + by, \; a,b \in R\}.$$

Define the linear functional $f:E \to R$ by

$$f(z) = a\|x\|^2 + b\langle y,x\rangle_s, \text{ for } z = ax + by \in E. \quad (2.14)$$

First of all we see that

$$f(x) = \|x\|^2, \; f(y) = \langle y,x\rangle_s. \quad (2.15)$$

Next we prove that f satisfies

$$f(ax + by) \leq \|x\| \|ax + by\|, \quad \text{for all } a,b \in R. \tag{2.16}$$

To simplify notation, set

$$\lambda_+ = \langle y,x\rangle_+, \ \lambda_- = \langle y,x\rangle_- \tag{2.16$'$}$$

and therefore to prove (2.16) it suffices to show that

$$a\|x\| + b\lambda_+ \leq \|ax + by\|, \quad \text{for all } a,b \in R. \tag{2.17}$$

The convexity of $t \to \|x + ty\|$ implies

$$\|x\| + t\lambda_+ \leq \|x + ty\|, \quad \text{for all } t \in R. \tag{2.18}$$

Assume now that $a > 0$. Then (2.18) gives

$$a\|x\| + b\lambda_+ = a(\|x\| + \frac{b}{a}\lambda_+) \leq a\|x + \frac{b}{a}y\| = \|ax + by\|.$$

If $a < 0$, $-a > 0$, so again using (2.18) we have

$$\|x\| + \frac{b}{-a}\lambda_+ \leq \|x - \frac{b}{a}y\|$$

which implies

$$-\|x\| + \frac{b}{-a}\lambda_+ \leq \|x - \frac{b}{a}y\| - 2\|x\| \leq \|-x - \frac{b}{a}y\| = -\frac{1}{a}\|ax + by\|.$$

Multiplying by $-a$ we obtain (2.17).

Finally, if $a = 0$, (2.17) becomes $b\lambda_+ \leq \|by\|$, which is a trivial consequence of $\lambda_+ \leq \|y\|$.

In this way (2.16) is proved. We now see that $\|f\| = \|x\|$ (by (2.15) and (2.16)). According to the Hahn-Banach theorem let w_i be an extension of f to the whole of X such that

$$w_1(x) = f(x) = \|x\|^2, \quad \|w_1\| = \|x\|.$$

In other words, $w_1 \in F(x)$ and by (2.15) we obtain $\langle y,x\rangle_s = f(y) = w_1(y)$.

Similarly we prove the existence of $w_2 \in F(x)$ such that $\langle y,x\rangle_i = w_2(y)$ and the proof is complete.

Definition 2.2 *Let* $\tilde{R} = [-\infty, +\infty]$. *The function* $f:X \to \tilde{R}$ *is said to be proper if it is nonidentically* $+\infty$ *and* $f(x) > -\infty$ *for all* $x \in X$. *The subset*

$$D_e(f) = \{x \in D(f),\ f(x) < +\infty\} \tag{2.19}$$

is called the effective domain of f.

Let C be a nonempty subset of X. The function

$$I_C(x) = \begin{cases} 0, & \text{if } x \in C \\ +\infty, & \text{if } x \in C \end{cases} \tag{2.20}$$

is said to be the *indicator function* of C. Clearly, C is convex iff I_C is convex, and $D_e(I_C) = C$.

Definition 2.3 *Let* X *be a real Banach space and* $f:X \to R$ *a proper function. The subdifferential* $\partial f(x)$ *of* f *at* x *is defined by*

$$\partial f(x) = \{x^* \in X^*,\ f(y) - f(x) \geqslant \langle x^*, y-x\rangle,\ \ \forall y \in X\}\ . \tag{2.21}$$

It is clear that $\partial f(x)$ may contain more than one element or may be the empty set. The domain $D(\partial f)$ of the subdifferential $\partial f:X \to 2^{X^*}$ is defined by

$$D(\partial f) = \{x \in X;\ \partial f(x) \text{ is nonempty}\}. \tag{2.22}$$

We now prove the following property of the duality mapping F.

Proposition 2.4 *The duality mapping* $F:X \to 2^{X^*}$ *is the subdifferential of the function* $f:X \to R_+$ *given by* $f(x) = 2^{-1}\,\|x\|^2$ (i.e. $\partial f(x) = F(x),\ \ \forall x \in X$).

Proof Let $x \in X$ and $x^* \in F(x)$. Then for each $y \in X$ we have

$$\frac{1}{2}\|x + y\|^2 - \frac{1}{2}\|x\|^2 = \|x + y\|\ \|x\| - \langle x^*, x\rangle \geqslant \langle x^*, y\rangle \tag{2.23}$$

because $\langle x^*, x\rangle = \|x\|^2$ and $\langle x^*, x + y\rangle \leqslant \|x\|\,\|x + y\|$. Hence $x^* \in f(x)$. Conversely, take $x^* \in \partial f(x)$, that is

$$\frac{1}{2}\|x + y\|^2 - \frac{1}{2}\|x\|^2 \geqslant \langle x^*, y\rangle,\ \ \forall y \in X. \tag{2.24}$$

Replacing y by ty with t > 0 and dividing by t, the inequality (2.24) yields

$$\langle x^*,y\rangle \leqslant \langle y,x\rangle_s \leqslant \|x\| \; \|y\|, \; \forall y \in X \tag{2.25}$$

which means that $\|x^*\| \leqslant \|x\|$. Now let us replace (in (2.24)) y = tx, with $t < 0$. Dividing by t and letting $t \uparrow 0$ we get $\langle x^*,x\rangle \geqslant \|x\|^2$. Therefore $\langle x^*,x\rangle = \|x\|^2$ and $\|x^*\| = \|x\|$, so $x^* \in F(x)$. The proof is complete.

Remark 2.1 It is well known that the subdifferential of a proper, lower-semicontinuous convex (l.s.c.) function is a maximal monotone operator. This implies (according to Proposition 2.4) that the duality mapping F of the Banach space X is a maximal monotone (possible multivalued) operator, i.e. $\langle x_1^* - x_2^*, x_1 - x_2\rangle \geqslant 0$, $\forall x_1,x_2 \in X$, $x_1^* \in F(x_1)$, $x_2^* \in F(x_2)$ and F is not properly contained in any other monotone set of $X \times X^*$.

Other important properties of F are given by:

Proposition 2.5 *For every* $x,y \in X$ *we have*

(a) $\langle y,x\rangle_s = \sup \{x^*(y);\ x^* \in F(x)\}$

(b) $\langle y,x\rangle_i = \inf \{x^*(y);\ x^* \in F(x)\}$

(c) *The following properties are equivalent:*

(i) *there exists* $x^* \in F(x)$ *such that* $x^*(y) \geqslant 0$;

(ii) $\langle y,x\rangle_s \geqslant 0$ (*or* $\langle x,y\rangle_+ \geqslant 0$);

(iii) $\|x\| \leqslant \|x + ty\|$, $\forall t > 0$.

Proof In view of (2.25), (a) is a consequence of Proposition 2.3. Similarly, one obtains (b) and the equivalence between (i) and (ii). According to (2.12) we see that the implication (iii) ⇒ (ii) is trivial. Finally, the convexity of $t \to \|x + ty\|$ shows that (ii) ⇒ (iii) and the proof is complete. □

Let us list some useful properties of the functions defined by (2.9) and (2.10)

Proposition 2.6 *Let* $x,y,z \in X$. *Then*

(1) $\langle y + z,x\rangle_p \leqslant \|y\| \; \|x\| + \langle z,x\rangle_p$, p = s *or* i;

(2) $\langle ay,bx\rangle_p = ab\langle y,x\rangle_p$, p = s *or* i, $a,b \in R$, $ab \geqslant 0$:

(3) $\langle ax + y,x\rangle_p = a\,\|x\|^2 + \langle y,x\rangle_p$, $p = s$ *or* i, $a \in R$;

(4) *if* $\langle x,y\rangle_s \leq \langle z,y\rangle_s$, *then* $\langle x - z,y\rangle_i \leq 0$;

(5) $\langle y, -x\rangle_s = \langle -y,x\rangle_s = -\langle y,x\rangle_i$, $\langle y,-x\rangle_i = \langle -y,x\rangle_i = -\langle y,x\rangle_s$;

(6) $\langle \cdot,\cdot\rangle_s : X \times X \to R$ *is upper semicontinuous.*

Proof Properties (2) and (5) follow directly from (2.9) and (2.10). For (1) and (3) we may combine Propositions 2.3 and 2.5. We now prove (4). Let $y_0^* \in F(y)$ be such that $\langle z,y\rangle_s = y_0^*(z)$. Then $y_0^*(x) \leq \langle x,y\rangle_s \leq y_0^*(z)$, therefore $y_0^*(x-z) \leq 0$, which implies $\langle x-z,y\rangle_i \leq 0$. It remains to prove (6), i.e. to show that for every $(x_n,y_n) \in X \times X$ with $x_n \to x$ and $y_n \to y$, we have

$$\limsup_{n \to \infty} \langle x_n,y_n\rangle_s \leq \langle x,y\rangle_s. \tag{2.26}$$

Set $\limsup_{n \to \infty} \langle x_n,y_n\rangle_s = c$. Since there is a subsequence of $\langle x_n,y_n\rangle_s$ convergent to c we may assume (for the simplicity of writing) that even $\lim_{n\to\infty} \langle x_n,y_n\rangle$ exists.

Let $y_n^* \in F(y_n)$ be such that $y_n^*(x_n) = \langle x_n,y_n\rangle_s$ (see Proposition 2.5).

Since $\|y_n^*\| = \|y_n\|$ it follows that y_n^* is bounded. Let y^* be any weak-star limit point of $\{y_n^*\}$. Then $y^* \in F(y)$ (see the proof of Proposition 2.1 (i)). Since $x_n \to x$ strongly and $y_n^* \rightharpoonup y^*$ weakly-star, it is easy to check that $y_n^*(x_n) \to y^*(x)$.

Using once again Proposition 2.5, we see that $y^*(x) \leq \langle x,y\rangle_s$, therefore $c = y^*(x) \leq \langle x,y\rangle_s$ so (2.26) is proved.

2.2 Duality mapping of some concrete spaces ($L^p, l^p, W_0^{1,p}, H_0^1$)

First of all, from Proposition 2.5 and (2.12) there immediately follows:

Corollary 2.1 *The norm of* X *is Gâteaux differentiable at any* $x \neq 0$ (i.e. $\lim_{t\to 0} \frac{\|x + ty\| - \|x\|}{t}$ *exists of all* $y \in X$) *if and only if the duality mapping* F *is single-valued. Moreover, if* $x \neq 0$, *then*

$$\lim_{t\to 0} \frac{\|x + ty\| - \|x\|}{t} = \langle \frac{1}{\|x\|}F(x),y\rangle = \frac{d}{dt}\,\|x+ty\|\Big|_{t=0},\ y \in X. \tag{2.27}$$

Thus we have derived the well-known result which asserts that if the norm of

X is Gâteaux differentiable at $x \neq 0$, then its Gateaux derivative is just $\frac{1}{\|x\|}F(x)$ (therefore it is a linear continuous functional, more precisely $y \to \langle\frac{1}{\|x\|} F(x), y\rangle$ is a linear continuous operator from X into R). In particular, if X* is strictly convex, then F is single-valued and therefore (2.27) holds. Recall that if X is reflexive under the norm $\|\cdot\|$, then there exists an equivalent norm $\|\cdot\|_1$ on X, such that X is strictly convex with respect to $\|\cdot\|_1$ and X* is strictly convex under the dual norm $\|\cdot\|_1^*$ (see Asplund [1]). Clearly (2.27) allows us to determine the duality mapping of some concrete space, whose norm is Gateaux differentiable (at $x \neq 0$). Let Ω be an open subset of R^n. Denote by $L^p(\Omega)$ with $p \geqslant 1$ the usual Banach space of all Lebesgue measurable functions $f:\Omega \to R$, with

$$\|f\|_p = \{\int_\Omega |f(x)|^p\, dx\}^{1/p} < +\infty, \quad 1 \leqslant p < +\infty \tag{2.28}$$

$$\|f\|_\infty = \operatorname*{essup}_{x \in \Omega} |f(x)|. \tag{2.29}$$

Proposition 1 *The duality mapping of* $L^p(\Omega)$ *is given by*

$$(Ff)(x) = \frac{f(x)|f(x)|^{p-2}}{\|f\|_p^{p-2}}, \quad f \in L^p(\Omega),$$

$$2 \leqslant p < \infty, \; x \in \quad , \; f \neq 0, \tag{2.30}$$

where $|f(x)|$ *denotes the absolute value of* $f(x) \in R$.

One can proceed to a direct verification that F given by (2.30) is just the duality mapping of L^p (taking into account the definition of F, the uniform convexity of $L^p(\Omega)$, $p > 1$ and the fact that the dual of $L^p(\Omega)$ is $L^q(\Omega)$ where $\frac{1}{p} + \frac{1}{q} = 1$). This is a trivial verification indeed. In what follows we show how to determine F. To find F, one computes the Gateaux derivative of norm of $L^p(\Omega)$ (this norm is Gateaux differentiable since $L^p(\Omega)$ is uniformly convex for $p > 1$). If $f \in L^p(\Omega)$, with $f \neq 0$, then for every $g \in L^p(\Omega)$ we have

$$\langle\frac{1}{\|f\|_p} F(f), g\rangle = \frac{d}{dt}\|f + tg\|_p\Big|_{t=0} = \frac{d}{dt}\{\int_\Omega |f(x) + tg(x)|^p dx\}^{1/p}\Big|_{t=0}$$

$$= \frac{1}{p}\{\int_\Omega |f(x)|^p\, dx\}^{\frac{1}{p}-1} \frac{d}{dt}\int_\Omega |f(x) + tg(x)|^p\, dx\Big|_{t=0}$$

$$= \|f\|_p^{(1-p)} \int_\Omega |f(x)|^{p-1}(\mathrm{sign}\ f(x))g(x)dx$$

(see Remark 2.2 below). Therefore

$$\langle F(f),g\rangle = \|f\|_p^{2-p} \int_\Omega |f(x)|^{p-2}f(x)g(x)dx,\ \forall g \in L^p(\Omega) \tag{2.31}$$

which proves (2.30).

Remark 2.2 In the proof above we have used the formula

$$\frac{d}{dt}\int_\Omega |f(x) + tg(x)|^p\, dx\Big|_{t=0} = \int_\Omega \frac{d}{dt}|f(x) + tg(x)|^p\Big|_{t=0} dx$$

$$= p\int_\Omega |f(x)|^{p-1}(\mathrm{sign}\ f(x))g(x)dx \tag{2.32}$$

as well as the elementary fact that if $f(x) \neq 0$, then

$$\frac{d}{dt}|f(x) + tg(x)| = (\mathrm{sign}\ f(x))g(x), |f(x)|\ (\mathrm{sign}\ f(x)) = f(x). \tag{2.33}$$

Let us observe that (2.32) is a consequence of the Lebesgue theorem.

Indeed, if we set $h(t) = |f(x) + tg(x)|^p$, $p > 1$, $f(x) \neq 0$, then the convexity of h implies

$$\frac{|f(x) + tg(x)|^p - |f(x)|^p}{t} = \frac{h(t) - h(0)}{t} \leq \frac{h(t_1) - h(0)}{t_1}$$

$$= \frac{|f(x) + t_1 g(x)|^p - |f(x)|^p}{t_1}$$

for all $t < t_1$, where $t_1 > 0$. Since

$$\lim_{t\to 0} \frac{|f(x) + tg(x)|^p - |f(x)|^p}{t} = \begin{cases} p|f(x)|^{p-1}(\mathrm{sign}\ f(x))g(x), & \text{if } f(x) \neq 0 \\ 0\ , & \text{if } f(x) = 0 \end{cases}$$

by the Lebesgue theorem one obtains (2.32). □

Set

$$l^p = \{x = (x_n)_{n\in\mathbb{N}},\ x_n \in R,\ \sum_{n=1}^{\infty} |x_n|^p < +\infty\}.$$

This is a Banach space with respect to the norm

$$\|x\| = \{\sum_{n=1}^{\infty} |x_n|^p\}^{1/p}. \tag{2.34}$$

Similarly, one proves that the duality mapping F of l^p is given by

$$F(x) = \|x\|^{2-p} (|x_n|^{p-2} x_n)_{n\in N}, \ x = (x_n)_{n\in N} \in l^p. \tag{2.35}$$

We now proceed to the determination of the duality mapping of the usual Sobolev space $W_o^{1,p}(\Omega)$ (see Chapter 5, §5). It is known that (in view of Poincaré inequality, see e.g. Pascali [1, p. 51]) an equivalent norm to the usual norm of $W_o^{1,p}(\Omega)$, is given by

$$\|f\|_{1,p} = \{\sum_{i=1}^{n} \int_\Omega |\frac{\partial f}{\partial x_i}|^p dx\}^{1/p}, \ f \in W_o^{1,p}(\Omega), \ p > 1. \tag{2.36}$$

Therefore, if $f \neq 0$, $f,g \in C_o^\infty(\Omega)$, by (2.27) and (2.32) we have

$$\langle \frac{1}{\|f\|_{1,p}} F(f),g\rangle = \frac{d}{dt} \|f + tg\|_{1,p}\Big|_{t=0} = \frac{d}{dt} \{ \sum_{i=1}^{n} \int_\Omega |\frac{\partial f}{\partial x_i}$$

$$+ t \frac{\partial g}{\partial x_i}|^p dx\}^{1/p}\Big|_{t=0} = \frac{1}{p} \{ \sum_1^n \int_\Omega |\frac{\partial f}{\partial x_i}|^p dx\}^{\frac{1}{p} - 1}$$

$$\sum_1^n \int_\Omega \frac{d}{dt} |\frac{\partial f}{\partial x_i} + t \frac{\partial g}{\partial x_i}|^p\Big|_{t=0} dx$$

$$= \|f\|_{1,p}^{1-p} \sum_{i=1}^{n} \int_\Omega |\frac{\partial f}{\partial x_i}|^{p-1} (\text{sign} \frac{\partial f}{\partial x_i}) \frac{\partial g}{\partial x_i} dx$$

$$= \|f\|_{1,p}^{1-p} \sum_1^n \int_\Omega |\frac{\partial f}{\partial x_i}|^{p-2} \frac{\partial f}{\partial x_i} \frac{\partial g}{\partial x_i} dx$$

$$= - \|f\|_{1,p}^{1-p} \int_\Omega \sum_{i=1}^{n} \frac{\partial}{\partial x_i} (|\frac{\partial f}{\partial x_i}(x)|^{p-2} \frac{\partial f}{\partial x_i}(x)) g(x)dx. \tag{2.37}$$

In view of the density of $C_0^\infty(\Omega)$ in $W_0^{1,p}(\Omega)$, it follows that the duality mapping of $W_0^{1,p}(\Omega)$ is given by

$$F(f) = -\|f\|_{1,p}^{2-p} \sum_{i=1}^{n} \frac{\partial}{\partial x_i}\left(\left|\frac{\partial f}{\partial x_i}\right|^{p-2} \frac{\partial f}{\partial x_i}\right), \; p \geqslant 2. \tag{2.38}$$

In the case p = 2, (2.38) becomes

$$F(f) = -\sum_{i=1}^{n} \frac{\partial^2 f}{\partial x_i^2} = -\Delta f. \tag{2.39}$$

In other words, the duality mapping of $W_0^{1,2}(\Omega) \equiv H_0^1(\Omega)$ is the Laplace operator.

2.3 Weak derivative and duality mapping

Let I be an open interval of the real axis R. A function $u: I \to X$ is said to be weakly differentiable at $t_0 \in I$ if, for each $x^* \in X^*$, the real valued function $t \to \langle x^*, u(t)\rangle$ is differentiable at $t = t_0$.

Denote by $u'(t_0)$ the weak derivative of u at t_0, that is

$$\lim_{h\to 0} \langle x^*, \frac{u(t_0+h) - u(t_0)}{h}\rangle = \langle x^*, u'(t_0)\rangle, \; \forall x^* \in X^*.$$

The following lemma of Kato is useful in the theory of evolution equations on general Banach spaces.

Lemma 2.1 (T. Kato). *Suppose that* $u: I \to X$ *is weakly differentiable at* $t_0 \in I$ *and* $t \to \|u(t)\|$ *is differentiable at* $t = t_0$. *Then*

$$\begin{aligned}\frac{1}{2}\frac{d}{dt}\|u(t)\|^2\Big|_{t=t_0} &= \|u(t_0)\| \frac{d}{dt}\|u(t)\|\Big|_{t=t_0} \\ &= \langle u'(t_0), u(t_0)\rangle_s = \langle u'(t_0), u(t_0)\rangle_i.\end{aligned} \tag{2.40}$$

Proof Let x^* be an arbitrary element of $F(u(t_0))$. Then

$$\begin{aligned}\langle x^*, u(t_0+h) - u(t_0)\rangle &= \langle x^*, u(t_0+h)\rangle - \|u(t_0)\|^2 \\ &\leqslant (\|u(t_0+h) - u(t_0)\|)\,\|u(t_0)\|.\end{aligned} \tag{2.41}$$

Dividing by $h > 0$ ($h < 0$) and then letting $h \to 0$ we get

$$\langle x^*, u'(t_0)\rangle = \|u(t_0)\| \frac{d}{dt} \|u(t)\| \Big|_{t=t_0}, \quad \forall x^* \in F(u(t_0)) \tag{2.42}$$

which implies (2.40) (see also (2.9) and (2.10)).

Remark 2.3 Let $x^* \in X^*$ and let $t \to \langle x^*, u(t)\rangle$ and $t \to \|u(t)\|$ be differentiable to the right of t_0.

If we assume that X^* is uniformly convex, then

$$\frac{1}{2} D^+ \|u(t)\|^2 \Big|_{t=t_0} = \langle D^+u(t_0), F(u(t_0))\rangle \tag{2.43}$$

where D^+ denotes the right derivative. A similar result holds by considering D^- (the left derivative) instead of D^+.

Proofof (2.43). Dividing (2.41) by $h > 0$ and letting $h \downarrow 0$, we get immediately

$$\langle F(u(t_0)), D^+u(t)\Big|_{t=t_0}\rangle \leq \frac{1}{2} D^+ \|u(t)\|^2 \Big|_{t=t_0}. \tag{2.44}$$

We now proceed to obtain the converse inequality. The following inequality holds:

$$\langle F(u(t_0 + h)), u(t_0 + h) - u(t_0)\rangle = \|u(t_0+h)\|^2 - \langle F(u(t_0+h)), u(t_0)\rangle$$

$$\geq \|u(t_0 + h)\| (\|u(t_0 + h)\| - \|u(t_0)\|).$$

Dividing by $h > 0$, using the continuity of the duality mapping and letting $h \downarrow 0$, one obtains the converse inequality of (2.44).

Thus (2.43) is proved. Clearly, the following fact has been used:

$$\lim_{h\downarrow 0} \langle F(u(t_0+h)), \frac{u(t_0+h) - u(t_0)}{h}\rangle = \lim_{h\downarrow 0} \langle F(u(t_0)),$$

$$\frac{u(t_0+h) - u(t_0)}{h}\rangle = \langle F(u(t_0)), D^+u(t)\Big|_{t=t_0}\rangle. \tag{2.45}$$

This fact is a consequence of the strong boundedness of $(u(t_0+h)-u(t_0))/h$ (which is weakly convergent as $h \downarrow 0$) and of the continuity of F. These remarks imply

$$\lim_{h\downarrow 0} \langle F(u(t_o+h)) - F(u(t_o)), \frac{u(t_o+h) - u(t_o)}{h}\rangle = 0,$$

which yields (2.45).

<u>Remark 2.4</u> The Formulas (2.40) and (2.43) can be written in the forms, respectively,

$$\|u(t_o)\| \frac{d}{dt} \|u(t)\| \Big|_{t=t_o} = \langle x^*, u(t_o)\rangle, \quad \forall x^* \in F(u(t_o)) \tag{2.46}$$

$$\|u(t_o)\| D^+ \|u(t)\| \Big|_{t=t_o} = \langle F(u(t_o)), D^+u(t)\Big|_{t=t_o}\rangle \tag{2.47}$$

(provided that all the involved derivatives exist).

§3. Dissipative (accretive) operators

3.1 Fundamental properties of dissipative (accretive) sets

In what follows the multivalued operators (sets) are not distinguished from their graphs. Therefore, a multivalued operator $A \cdot D(A) \subset X \to 2^X$ is regarded as a subset of $X \times X$. More precisely, if $A \subset X \times X$, we define

$$\left.\begin{array}{ll} \text{(a)} & Ax = \{y \in X;\ [x,y] \in A\} \\ \text{(b)} & D(A) = \{x \in X;\ Ax \neq \emptyset\} \\ \text{(c)} & R(A) = \bigcup_{x\in D(A)} Ax \end{array}\right\} \tag{3.1}$$

Thus $[x,y] \in A$ denotes the fact that $x \in D(A)$ and $y \in Ax$.

If $A,B \subset X \times X$ and a is a real number, one defines

$$\left.\begin{array}{ll} \text{(a)} & A + B = \{[x,y + z];\ x \in D(A) \cap D(B),\ y \in Ax \text{ and } z \in Bx\} \\ \text{(b)} & aA = \{[x,ay];\ y \in Ax\} \\ \text{(c)} & A^{-1} = \{[y,x];\ [x,y] \in A\}. \end{array}\right\}$$

Obviously, when A is a single-valued operator the subset Ax defined in (3.1)(a) denotes the value of A at x.

<u>Definition 3.1</u> *A subset* $A \subset X \times X$ *is said to be accretive if for every* $x_1,x_2 \in D(A)$, *there is* $w \in F(x_1-x_2)$ *such that:*

$$\langle w, y_1 - y_2 \rangle \geq 0, \textit{ for all } y_i \in Ax_i, \ i = 1,2. \tag{3.2}$$

A *is said to be dissipative if* -A *is accretive.*

Proposition 3.1 *The accretiveness of the subset* A *is equivalent to each of Conditions* (i) *and* (ii) *below:*

(i) $\|x_1 - x_2\| \leq \|x_1 - x_2 + t(y_1 - y_2)\|, \ \forall [x_i, y_i] \in A, \ t > 0$;

(ii) $\langle y_1 - y_2, x_1 - x_2 \rangle_s \geq 0, \forall [x_i, y_i] \in A, \ i = 1,2.$

Proof Each of Conditions (i) and (ii) is equivalent to the Condition (3.2) (in view of Propositions 2.3 and 2.5). □

Taking into account that $\langle -y, x \rangle_s = - \langle y, x \rangle_i$ (Proposition 2.6) in the dissipative case the result corresponding to Proposition 3.1 is the following one:

Proposition 3.2 *The subset* $A \subset X \times X$ *is dissipative (i.e. for every* $x_i \in D(A)$, *there is* $w \in F(x_1 - x_2)$ *such that*

$$\langle w, y_1 - y_2 \rangle \leq 0 \textit{ for all } [x_i, y_i] \in A, \ i = 1,2$$

if and only if

(i) $\|x_1 - x_2\| \leq \|x_1 - x_2 - t(y_1 - y_2)\|, \ \forall t \geq 0, \ [x_i, y_i] \in A, \ i = 1,2$

or

(ii) $\langle y_1 - y_2, x_1 - x_2 \rangle_i \leq 0, \ \ \forall [x_j, y_j] \in A, \ j = 1,2.$

In what follows we assume that $A - \omega I$ is dissipative, where $A \subset X \times X$ and ω is a real number. In view of Proposition 3.2, this is equivalent to

(1) $\|x_1 - x_2\| \leq \|(x_1 - x_2)(1 + t\omega) - t(y_1 - y_2)\|$

or (3.3)

(2) $\langle y_1 - y_2, x_1 - x_2 \rangle_i \leq \omega \| x_1 - x_2 \|^2$

for all $t > 0$ and $[x_i, y_i] \in A$, $i = 1,2$. Take $\lambda > 0$ such that $\lambda\omega < 1$. Then, for $t = \lambda/(1 - \lambda\omega)$, (3.3) yields (3.4) and conversely (see Remark 3.2):

$$\|x_1 - x_2\| (1 - \lambda\omega) \leq \|x_1 - x_2 - \lambda(y_1 - y_2)\|$$

$$= \|x_1 - \lambda y_1 - (x_2 - \lambda y_2\| \qquad (3.4)$$

for all $\lambda > 0$ with $\lambda\omega < 1$, and $[x_i, y_i] \in A$, $i = 1,2$.

Let $u \in R(I - \lambda A)$. It follows from (3.4) that there exist a unique $x_u \in D(A)$ and a unique $y_u \in Ax_u$ such that $u = x_u - \lambda y_u$. Indeed, if we have also $u = x - \lambda y$ for some $[x,y] \in A$, then by (3.4) we obtain $x_u = x$ (and consequently $y_u = y$). It is natural to denote $x_u = (1 - \lambda A)^{-1}u \equiv J_\lambda u$. Set also $y_u = A_\lambda u$. Therefore,

$$\left.\begin{aligned} &J_\lambda : R(I - \lambda A) \to D(A), \quad J_\lambda(x - \lambda y) = x, \; A_\lambda(x - \lambda y) = y \\ &\quad u = J_\lambda u - \lambda A_\lambda u. \end{aligned}\right\} \qquad (3.5)$$

for each $\lambda > 0$, $[x,y] \in A$ and $u \in R(I - \lambda A)$.

In the next lemma, the main properties of the operator J_λ are presented.

Lemma 3.1 *Let* $\omega \in R$ *and* $A \subset X \times X$ *be such that* $A-\omega I$ *is dissipative. Then for each* $\lambda > 0$ *with* $\lambda\omega < 1$, $J_\lambda : R(I - \lambda A) \to D(A)$ *defined by* (3.5) *has the properties*

(a) $\|J_\lambda u - J_\lambda v\| \leq (1 - \lambda\omega)^{-1} \|u-v\|, \forall [u,v] \in R(I - \lambda A)$;

(b) $\|J_\lambda u - u\| \leq \lambda(1 - \lambda\omega)^{-1} |Au|, \forall u \in D(A) \cap R(I - \lambda A)$

where

$$|Au| = \inf \{ \|y\| \, ; \, y \in Au\}; \qquad (3.6)$$

(c) *if* $D(A) \subset R(I - \lambda A)$ *and* n *is a natural number, then*

$$\|J_\lambda^n u - u\| \leq n(1 - \lambda|\omega|)^{-n+1} \|J_\lambda u - u\| \leq n\lambda(1 - \lambda|\omega|)^{-n}|Au|$$

$$\lambda|\omega| < 1, u \in D(A);$$

(d) *if* $u \in R(I - \lambda A)$, $\mu > 0$ *with* $\mu\omega < 1$, *then the nonlinear version of "resolvent formula" holds, i.e.*

$$\frac{\mu}{\lambda} u + \frac{\lambda-\mu}{\lambda} J_\lambda u \in R(I - \mu A), \text{ and } J_\lambda u = J_\mu\left(\frac{\mu}{\lambda} u + \frac{\lambda-\mu}{\lambda} J_\lambda u\right).$$

Proof Let $[x_i,y_i] \in A$, $i = 1,2$, be such that $u = x_i - \lambda y_1$ and $v = x_2 - \lambda y_2$. Then, by (3.5), $J_\lambda u = x_1$, $J_\lambda v = x_2$, hence (a) is a direct consequence of (3.4).

To prove (b) one observes that if $u \in D(A)$, then $u = J_\lambda(u - \lambda y)$, $\forall y \in Au$; therefore, making use of (a), we get

$$\|J_\lambda u-u\| = \|J_\lambda u-J_\lambda(u - \lambda y)\| \leq \lambda(1 - \lambda\omega)^{-1} \|y\|, \quad \forall y \in Au,$$

which implies (b).

In order to obtain (c) one uses triangle inequality (in standard manner):

$$\|J_\lambda^n u - u\| \leq \sum_{i=0}^{n-1} \|J_\lambda^{n-i} u_\lambda - J^{n-(i+1)} u\| \leq \sum_{i=0}^{n-1} (1-\lambda\omega)^{-n+i+1} \|J_\lambda x-x\| . \tag{3.7}$$

We now observe that $\lambda|\omega| < 1$ implies $(1 - \lambda\omega)^{-1} \leq (1 - \lambda|\omega|)^{-1}$, hence

$$\sum_{i=0}^{n-1} (1 -\lambda\omega)^{-n+i+1} \leq n(1 - \lambda|\omega|)^{-n+1}$$

and thus (c) follows. Finally let us prove (d). There is $[x,y] \in A$ such that $u = x - \lambda y$, therefore $J_\lambda u = x$. Consequently

$$\frac{\mu}{\lambda} u + \frac{\lambda-\mu}{\lambda} J_\lambda u = x - \mu y \in R(I-\mu A), \; J_\mu(x-\mu y) = x = J_\lambda u$$

and the proof is complete.

By (3.5), the connection between A_λ and J_λ is given by

$$A_\lambda = \frac{J_\lambda - I}{\lambda} = \frac{(I - \lambda A)^{-1} - I}{\lambda}, \; \lambda > 0 \tag{3.8}$$

where $A-\omega I$ is dissipative.

In the case $A + \omega I$ - accretive, J_λ and A_λ have the form

$$J_\lambda = (I + \lambda A)^{-1}, \quad A_\lambda = \lambda^{-1}(I - J_\lambda). \tag{3.9}$$

Lemma 3.2 *In the conditions of Lemma 3.1, A_λ has the properties:*

(a) $(1-\lambda\omega)\| A_\lambda u\| \leq (1-\mu\omega) \|A_\mu u\|$, $\lambda\omega < 1$, $\omega < 1$, $0 < \mu \leq \lambda$,

$u \in R(I - \lambda A) \cap R(I-\mu A)$;

(b) $\|A_\lambda u - A_\lambda v\| \leq \lambda^{-1}[1 + (1-\lambda\omega)^{-1}] \|u-v\|$, $u,v \in R(I - \lambda A)$;

(c) *if* $u \in R(I-\lambda A)$ *then* $A_\lambda u \in AJ_\lambda u$, *and* $\|A_\lambda u\| \leq (1-\lambda\omega)^{-1}|Au|$;

(d) $\langle A_\lambda u - A_\lambda v, u-v\rangle_s \leq \frac{\omega}{1-\lambda\omega} \|u-v\|^2$, $\forall u,v \in R(I-\lambda A)$.

Proof By definition of A_λ and Lemma 2.1 we have successively

$$\|A_\lambda u\| = \lambda^{-1} \|J_\lambda u - u\| \leq \lambda^{-1} \|J_\mu u - u\| + \lambda^{-1} \|J_\mu u - J_\lambda u\|$$

$$\leq \mu\lambda^{-1} \|A_\lambda u\| + \lambda^{-1} \|J_\lambda(\frac{\lambda}{\mu} u + \frac{\mu-\lambda}{\mu} J_\mu u) - J_\lambda u\| \leq \mu\lambda^{-1} \|A_\mu u\|$$

$$+ (\lambda-\mu)(1-\lambda\omega)^{-1} \lambda^{-1} \|A_\mu u\|$$

which yields (a). Property (b) follows from (3.8) and Lemma 3.1(a). We now prove (c). By hypothesis, $u = x - \lambda y$, with $[x,y] \in A$, therefore $x = J_\lambda u$, $y = A_\lambda u$. Thus $y \in A x$ means $A_\lambda u \in AJ_\lambda u$.

By Lemma 3.1(b) we see that

$$\|A_\lambda u\| = \lambda^{-1} \|J_\lambda u - u\| \leq (1-\lambda\omega)^{-1} |Au|. \tag{3.10}$$

(d) By Proposition 2.3 there is $x^* \in F(u-v)$, such that

$$\langle A_\lambda u - A_\lambda v, u-v\rangle_s = \langle A_\lambda u - A_\lambda v, x^*\rangle = \frac{1}{\lambda} \langle J_\lambda u - J_\lambda v, x^*\rangle - \frac{1}{\lambda} \langle u-v, x^*\rangle.$$

Since $\|x^*\| = \|u-v\|$ and $\langle u-v, x^*\rangle = \|u-v\|^2$, the result follows.

3.2 Maximal dissipative sets

Definition 3.2 (i) *A dissipative subset* $A \subset X \times X$ *is said to be maximal dissipative if it is not properly contained in any other dissipative subset of* $X \times X$.

(ii) *The dissipative subset* A *is said to be* m-dissipative *(or hypermaximal dissipative) if*

$$R(I - A) = X. \tag{3.11}$$

(iii) A *is said to be maximal accretive if* $-A$ *is maximal dissipative.*

Remark 3.1 Obviously a dissipative subset $A \subset X \times X$ is maximal dissipative iff whenever $[x_1,y_1] \in X \times X$ satisfies

$$\langle y - y_1, x - x_1 \rangle_i \leqslant 0, \quad \forall[x,y] \in A, \tag{3.12}$$

we obtain $[x_1,y_1] \in A$.

Proposition 3.3 $A \subset X \times X$ *is m-dissipative iff*

$$R(I - \lambda A) = X, \forall\lambda > 0. \tag{3.13}$$

Proof In this case $J_1 = (I-A)^{-1}$ is defined on the whole of X and, from Lemma 3.1(a) with $\omega = 0$, it follows that

$$\|(I-A)^{-1}u - (I-A)^{-1} v\| \leqslant \|u-v\|, \quad \forall u,v \in X. \tag{3.14}$$

Actually we have to prove that

$$R(I-A) = X \Rightarrow R(I-\lambda A) = X, \forall\lambda > 0 \tag{3.15}$$

in Condition (3.14).

Take an arbitrary $y \in X$. We must show that there is $x \in D(A)$ and $z \in Ax$ such that $y = x - \lambda z$, or equivalently

$$x = (I-A)^{-1}[\lambda^{-1}y + \lambda^{-1}(\lambda-1)x].$$

Define $T:X \to D(A)$, by

$$Tx = (I-A)^{-1}[\lambda^{-1}y + \lambda^{-1} (\lambda-1)x].$$

Then (3.14) implies that T is Lipschitz continuous of Lipschitz constant $|\lambda^{-1}-1|$. Therefore, for each $\lambda > \frac{1}{2}$, $|\lambda^{-1} -1| < 1$ so T has a fixed point.

Thus we have proved (3.15) for each $\lambda > \frac{1}{2}$.

Since λA is also dissipative (for any $\lambda > 0$), we can apply the previous result to $\lambda A, \lambda^2 A ,\ldots,\lambda^n A,\ldots$.

In other words, we have proved that

$$R(I-A) = X \Rightarrow R(I -\lambda^n A) = X, \forall n \in N \text{ and } \lambda > \frac{1}{2}$$

which implies (3.15).

Proposition 3.4 *Any m-dissipative subset $A \subset X \times X$ is maximal dissipative. (The converse statement is not necessarily true, unless X is a Hilbert space.)*

Proof We have to prove that if

$$\langle y-y_1, x-x_1 \rangle_i \leq 0, \quad \forall [x,y] \in A \tag{3.16}$$

then $[x_1,y_1] \in A$ (see (3.12)). Since $x_1-y_1 \in R(I-A) = X$, there is $[x,y] \in A$ such that

$$x_1-y_1 = x-y, \text{ or } y-y_1 = x - x_1.$$

Then (3.16) yields $\langle x - x_1, x - x_1 \rangle_i = \|x-x_1\|^2 \leq 0$, therefore $x = x_1$ and $y = y_1$.

Theorem 3.1 (1) *Let $A \subset X \times X$ be maximal dissipative. Then*

(1) *A is closed (hence the subset Ax is closed for each $x \in D(A)$).*

(2) *If $\{x_\lambda\} \subset D(A)$ has the properties: $x_\lambda \to x$, $A_\lambda x_\lambda \to y$ as $\lambda \downarrow 0$, then $[x,y] \in A$. For every $z \in \overline{D(A)}$, $J_\lambda z \to z$.*

(3) *If X* is strictly convex then Ax is closed and convex (for each $x \in D(A)$).*

(4) *If X is uniformly convex then*

(i) *A is demiclosed (in the sense of Definition 2.1);*

(ii) *given $\{x_\lambda\} \subset X$ such that $x_\lambda \to x$ as $\lambda \downarrow 0$ and $A_\lambda x_\lambda$ is bounded, then $x \in D(A)$ and any weak cluster point y of $A_\lambda x_\lambda$ belongs to Ax (i.e. $y \in Ax$).*

(5) *Assume that A is m-dissipative and X* is uniformly convex. Then*

(a) $\|A_\lambda x_\lambda\| \uparrow |Ax|$ *as $\lambda \downarrow 0$ for each $x \in D(A)$ (see (3.6));*

(b) *If both X and X* are uniformly convex, then for each $x \in D(A)$ there is a unique element $A_0 x \in Ax$ of minimal norm, i.e.*

$$\|A_0 x\| = |Ax| = \inf \{ \|y\| ; y \in Ax\} \tag{3.17}$$

and $\{A_\lambda x\}$ contains a subsequence which is convergent to $A_0 x$.

In addition (if X and X are uniformly convex),*

(c) *$\overline{D(A)}$ is a convex subset of X.*

Proof (1) Let $[x_n,y_n] \in A$ with $x_n \to x_0$ and $y_n \to y_0$ as $n \to \infty$. Since A is dissipative, we have (Proposition 3.2)

$$\|x_n-x\| \leqslant \|x_n - x - t(y_n-y)\| , \quad \forall t \geqslant 0, [x,y] \in A.$$

Letting $n \to \infty$, one obtains $\|x_0-x\| \leqslant \|x_0-x-t(y_0-y)\|$, $\forall t \geqslant 0$ and $[x,y] \in A$ or equivalently

$$\langle y-y_0, x-x_0\rangle_i \leqslant 0, \forall [x,y] \in A$$

which implies $[x_0,y_0] \in A$.

(2) From the hypothesis it follows that $A_\lambda x_\lambda$ is bounded. By (3.8), $J_\lambda x_\lambda - x_\lambda = \lambda A_\lambda x_\lambda$, therefore $J_\lambda x_\lambda \to x$ as $\lambda \downarrow 0$. In view of Lemma 3.2(c), $[J_\lambda x_\lambda, A_\lambda x_\lambda] \in A$ and therefore $[x,y] \in A$. If $x \in D(A)$, then Lemma 3.1(b) shows that $\lim_{\lambda \downarrow 0} J_\lambda x = x$. Making use of the Lipschitz continuity of J_λ (Lemma 2.1(a) with $\omega = 0$) we easily prove the above property for every $x \in \overline{D(A)}$.

(3) We have already proved that if $x_0 \in D(A)$ then $A x_0$ is closed. In order to show the convexity of Ax_0, let $y_1, y_2 \in Ax_0$ and $\lambda \in [0,1]$. Then

$$\langle y - [\lambda y_1 + (1 - \lambda)y_2], F(x - x_0)\rangle = \lambda\langle y - y_1, F(x - x_0\rangle$$

$$+ (1 -\lambda) \langle y - y_2, F(x - x_0)\rangle \leqslant 0, \quad \forall [x,y] \in A$$

which implies $\lambda y + (1 - \lambda)y_2 \in Ax_0$.

(4) (i) Let $[x_n,y_n] \in A$ with $x_n \to x_0$ (strongly) and $y_n \rightharpoonup y_1$ (weakly) as $n \to \infty$. Since the duality mapping F is continuous, letting $n \to \infty$ in the inequality

$$\langle y_n - y, F(x_n - x)\rangle \leqslant 0, \forall [x,y] \in A$$

we get

$$\langle y_0 - y, F(x_0 - x)\rangle \leqslant 0, \forall [x,y] \in A$$

therefore $[x_0,y_0] \in A$.

(ii) From the proof of (2) we see that $J_\lambda x_\lambda \to x$ as $\lambda \downarrow 0$. Now, since $[J_\lambda x_\lambda, A_\lambda x_\lambda] \in A$, (i) implies $[x,y] \in A$.

(5) (a) By Lemma 3.2 (with $\omega = 0$) we have

$$|Ax| \geq \|A_\lambda x\| \geq \|A_\mu x\|, \quad 0 < \lambda \leq \mu, \quad x \in D(A)$$

hence $\lim\limits_{\lambda \downarrow 0} \|A_\lambda x\| \equiv t$ exists and $t \leq |Ax|$.

Since X is reflexive we may assume that even $A_\lambda x \rightharpoonup y$ as $\lambda \downarrow 0$. By (4), $y \in Ax$, therefore $|Ax| \leq \|y\|$. In the case $y = 0$,

(a) is trivially satisfied so let us consider $y \neq 0$. Using

$$\langle A_\lambda x, F(y)\rangle \leq \|A_\lambda x\| \; \|y\| \text{ and } \lim_{\lambda \downarrow 0} \langle A_\lambda x, F(y)\rangle = \|y\|^2$$

one obtains

$$\|y\| \leq \lim_{\lambda \downarrow 0} \|A_\lambda x\| = t$$

and (a) follows.

(b) The existence of $A_0 x$ is known (see Corollary 1.1). The fact that $A_\lambda x$ (more precisely, a subsequence of $A_\lambda x$) is strongly convergent to $A_0 x$ as $\lambda \downarrow 0$ is a consequence of Lemma 1.1.

(c) Take $x_1, x_2 \in \overline{D(A)}$, $\alpha \in]0,1[$ and set

$$x_\alpha = \alpha x_1 + (1-\alpha)x_2, \; r_1(\lambda) = \|x_\alpha - x_1\| + \lambda \|A_0 x_1\|, \; r_2(\lambda)$$
$$= \|x_\alpha - x_2\| + \lambda \|A_0 x_2\|.$$

Obviously, by Lemma 3.1(b) (with $\omega = 0$)

$$\|J_\lambda x_\alpha - x_1\| \leq r_1(\lambda), \quad \|J_\lambda x_\alpha - x_2\| \leq r_2(\lambda), \quad r_1(\lambda) \to \|x_\alpha - x_1\|$$
$$r_2(\lambda) \to \|x_\alpha - x_2\| \text{ as } \lambda \downarrow 0,$$

hence

$$J_\lambda x_\alpha \in B(x_1, r_1(\lambda)) \cap B(x_2, r_2(\lambda)) \equiv C(\lambda), \quad \forall \lambda > 0.$$

In view of Proposition 1.2 we see that $\lim\limits_{\lambda \downarrow 0} J_\lambda x_\alpha = x_\alpha$ and, since $J_\lambda x_\alpha \in D(A)$, it follows that $x_\alpha \in \overline{D(A)}$. □

Remark 3.2 We now give a simple proof of the fact that $A \subset X \times X$ is ω-dissipative *iff* (3.4) holds. Indeed, if A is ω-dissipative then we have (3.3) and therefore there exists $x^* \in F(x_1-x_2)$ such that

$$x^*(y_1-y_2) \leq \omega \|x_1 - x_2\|^2 .$$

Consequently

$$\|x_1-x_2\|^2 = x^*(x_1-x_2) = x^*(x_1-x_2-\lambda(y_1-y_2)) + \lambda x^*(y_1-y_2)$$

$$\leq \|x_1-x_2\| \; (\|x_1-x_2 - \lambda(y_1-y_2)\|) + \lambda \, \omega \|x_1-x_2\|),$$

which implies (3.4). Conversely, (3.4) can be written in the form

$$-\lambda^{-1}(\|x_1-x_2-\lambda(y_1-y_2)\| - \|x_1-x_2\|) \leq \omega \|x_1-x_2\|.$$

Multiplying by $\|x_1-x_2\|$ and then letting $\lambda \downarrow 0$, one obtains (see (2.12))

$$\langle y_1-y_2, x_1-x_2\rangle_i \leq \omega\|x_1-x_2\|^2$$

which means that A is ω-dissipative.

We end this chapter by the following remark. If an operator $A:D(A) \subset X \to X$ is Lipschitz continuous, i.e.

$$\|Ax - Ay\| \leq L \|x - y\| , \quad \forall x,y \in D(A)$$

then A is obviously L-dissipative, i.e.

$$\langle Ax - Ay, x - y\rangle_i \leq L \|x-y\|^2 , \quad \forall x,y \in D(A)$$

since, by (2.12), $\langle Ax-Ay, x-y\rangle_i \leq \|Ax-Ay\| \, \|x-y\|$.

The converse implication is not true, even in the case X = R. For example, $A_0(x) = - \sqrt{x}$, $x \geq 0$, is clearly L-dissipative, but not Lipschitz continuous. The most important operators that are dissipative but not continuous are the dissipative partial differential operators (e.g. Laplace operator in Chapter 5).

§4. Notes and Remarks

Uniformly convex spaces were introduced by Clarkson in 1936. The proof of Proposition 2.2 is taken from Browder [2]. The proof of Proposition 2.3 follows Benilan [1] while that of Proposition 2.6 can be found in Crandall and Liggett [1]. Note that Properties (a) and (b) in Proposition 2.5 are particular cases of the following result.

Theorem 4.1 *Let* X *be a Banach space and let* $f:X \to]-\infty, +\infty]$ *be a convex proper function. If* f *is continuous at* $x \in \mathring{D}_e(f)$ *(the interior of* $D_e(f)$ *given by* (2.19)), *then for every* $y \in X$,

$$f'_+(x,y) = \sup \{x^*(y);\ x^* \in \partial f(x)\}$$

$$f'_-(x,y) = \inf \{x^*(y);\ x^* \in \partial f(x)\}.$$

Here $f'_+(x,y)$ denotes the directional derivative of f at x, to the right, evaluated at y, i.e. $f'_+(x,y) = \lim_{t \downarrow 0} t^{-1}(f(x+ty) - f(x))$. Similarly, $f'_-(x,\cdot)$ is the derivative of f at x, to the left. In the particular case in which $f(x) = 2^{-1}\,||x||^2$ we have $f'_+(x,y) = \langle y,x\rangle_s$, $f'_-(x,y) = \langle y,x\rangle_i$ and (by Proposition 2.4) $\partial f(x) = F(x)$. Therefore, in this case Theorem 4.1 leads to Properties (a) and (b) of Proposition 2.5.

There exists an extensive literature on the topics dealt with in this chapter. In this connection we refer to Barbu [2], Brezis [3], Browder [2], Da Prato [1], Kachurovskii [1]. Kobayashi [1], Minty [1], Rockafellar [1], Vainberg [1], etc.

2 Differential equations on closed subsets – flow invariance

§1. Peano's Theorem is a characterization of finite dimensionality (Godunov's Theorem)

Let us consider the Cauchy Problem (frequently denoted by $(CP;x_0)$)

$$x'(t) = F(t,x(t)), \quad x(t_0) = x_0, \quad x_0 \in X, \tag{1.1}$$

where $F:R \times X \to X$ is a continuous function. If X is a Banach space of *finite* dimension, then classical Peano's existence theorem asserts that the continuity of F ensures local existence of the solution to (1.1). It was shown by Dieudonné (1950) that if X is a Banach space of *infinite* dimension, then the continuity no longer ensures this local existence. The evolution of research in this direction is discussed in Section 6. Here we give:

Theorem 1.1 (Godunov). *Let X be a Banach space of infinite dimension. Then there exists a continuous function $F:R \times X \to X$ such that the Cauchy Problem*

$$x'(t) = F(t,x(t)), \quad x(0) = 0 \tag{1.2}$$

has no (strong) solution on any nonvanishing interval containing the origin.

The proof of this theorem is essentially based on the theorem of M. Day [1].

Theorem 1.2 (M. Day). *Let X be an infinite dimensional space. Then there exist biorthogonal sequences $\{e_n, n = 1,2,...\} \subset X$, $\{e_n^*, n = 1,2,...\} \subset X^*$ such that:*

(i) *$\{e_n\}$ is a Schauder basis for the closed linear manifold L in X spanned by the set of all e_n;*

(ii) $\|e_n\| = \|e_n^*\| = 1$, $e_n^*(e_m) = 0$ *for* $m \neq n$, $m,n = 1,2,...;$

(iii) *Setting $P_m(x) = \sum_{i=1}^{m} e_i^*(x)e_i$ for each $m = 1,2,...$, the linear operator P_m is a projection in L, of norm $\|P_m\| \leqslant 1 + 1/m$.*

Proof of Theorem 1.1 Let us consider the sequences $a_n = 1/(2n+1)$, $b_n = 1/2n$ and $c_n = (a_n + b_n)/2$.

Let ϕ_n, ψ_n and h be continuous real functions defined by $\phi_n(t) = 0$ outside $]a_n,c_n[$, $0 < \phi_n(t) < \frac{1}{n}$ if $t \in]a_n,c_n[$,

$$\psi_n(t) = \begin{cases} 0, \text{ if } t \leq c_n \\ \dfrac{t-c_n}{b_n-c_n}, \text{ if } t \in [c_n,b_n]; \\ 1, \text{ if } t \geq b_n \end{cases} \qquad h(t) = \begin{cases} 0, \text{ if } t < 0 \\ t, \text{ if } t \in [0,1] \\ 1, \text{ if } t \geq 1 \end{cases} \tag{1.3}$$

We now define $P:R \times L \to L$ and $\phi:R \times L \to L$ by

$$P(t,x) = \sum_{n=1}^{\infty} \psi_n(t)e_n^*(x)e_n,$$

$$\phi(t,x) = \sum_{n=1}^{\infty} \phi_n(t)h\left(\frac{(t-b_{n+1})^2}{4} - \|x-P_n(x)\|\right)e_n. \tag{1.4}$$

It is obvious that P and ϕ are continuous, hence the function $f:R \times L \to L$ below is also continuous:

$$f(t,x) = \begin{cases} \dfrac{P(t,x)}{\sqrt{\|P(t,x)\|}} + \phi(t,x), & \text{if } P(t,x) \neq 0 \\ \phi(t,x) & , \text{if } P(t,x) = 0. \end{cases} \tag{1.5}$$

On the basis of the Dugundji theorem [1, p.357], f admits a continuous extension $F:R \times X \to L$. We shall prove that such a continuous extension F of f satisfies the required property. In other words, let us prove that with such F, the conclusion of the theorem holds. Assume by contradiction that the problem (1.2) admits a solution $x:V \to X$, defined on a nonvanishing interval V containing the origin. Since

$$x(t) = \int_0^t F(s,x(s))ds \text{ and } F(s,x(s)) \in L,$$

it follows that actually $x(t) \in L$ for all $t \in V$. Consequently, x is a solution of the problem

$$x'(t) = f(t,x(t)), \quad x(0) = 0, \quad t \in V. \tag{1.6}$$

We see that $x = 0$ is not a solution of (1.6).

It is clear that only the following two situations may occur:

(I) There exist sufficiently large natural numbers n such that $x(b_n) \neq 0 (b_n \in V)$.

(II) $x(b_n) = 0$ for all sufficiently large n.

Let us now proceed to prove that each of the above situations yields a contradiction. We begin with (I). Fix m such that $x(b_m) \neq 0$. First of all we prove that, for all $n < m$,

$$e_n^*(x(b_m)) = \int_0^{b_m} e_n^*(f(s,x(s)))ds = 0 \tag{1.7}$$

We obviously have

$$e_n^*(P(s,x(s))) = \psi_n(s) e_n^*(x(s)) = 0 \tag{1.8}$$

$$e_n^*(\phi(s,x(s))) = \phi_n(s) h\left(\frac{(s-b_{n+1})^2}{4} - \|x(s)-P_n(x(s)\|\right) = 0 \tag{1.9}$$

for $s < b_m$ and $n < m$ (since in this case $\psi_n(s) = \phi_n(s) = 0$).

Combining (1.7) and (1.8), we get

$$e_n^*(f(s,x(s))) = 0, \quad s \leqslant b_m, \quad n < m, \tag{1.10}$$

which implies (1.7). Since $x(b_m) \neq 0$, the function u below is well defined:

$$u(t) = \frac{x(b_m)}{\|x(b_m)\|} \quad \frac{(t-\tilde{b}_m)^2}{4} \tag{1.11}$$

where $\tilde{b}_m$ satisfies

$$\tilde{b}_m < b_m, \ (b_m - \tilde{b}_m)^2 = 4\,\|x(b_m)\|. \tag{1.12}$$

We now prove that u is the unique solution to the problem

$$u'(t) = f(t,u(t)), \quad u(b_m) = x(b_m), \quad t \geqslant b_m. \tag{1.13}$$

To do this, let us note that, by (1.7), $e_n^*(u(t)) = 0$ for $t \geqslant b_m$ and $n < m$; hence

$$P_n(u(t)) = 0, \quad \forall n < m, \quad t \geqslant b_m. \tag{1.14}$$

Using (1.14) one verifies that, for all $t \geq b_m$,

$$\phi(t,u(t)) = \sum_{n=1}^{m-1} \phi_n(t)h(\frac{(t-b_{n+1})^2}{4} - \frac{(t-\tilde{b}_m)^2}{4})e_n = 0 \tag{1.15}$$

Moreover, for $t \geq b_m$ and $k \geq m$, $\psi_k(t) = 1$. This remark, and the fact that $e_n^*(u(t)) = 0$ if $n < m$, yield

$$P(t,u(t)) = \sum_{k=m}^{\infty} e_k^*(u(t))e_k = \sum_{k=1}^{\infty} e_k^*(u(t))e_k = u(t). \tag{1.16}$$

It is now clear that (in view of (1.15) and (1.16)) u is a solution to (1.13). The uniqueness of the solution to (1.13) is a consequence of the fact that f is locally Lipschitz in a neighbourhood of M, where

$$M = \{(t,u(t)),\ t \in [b_m,1]\} \subset R \times L.$$

To show this property of f, one verifies that $\phi(t,x) = 0$ in a neighbourhood of M, while $P(t,x) \neq 0$. Moreover, for $t \in [b_{n+1},b_n]$ we have

$$\|P(t,x) - P(t,y)\| = \|\sum_{k=1}^{\infty} \psi_k(t)e_k^*(x-y)e_k\|$$

$$= \|\psi_n(t)e_n^*(x-y)e_n + \sum_{k=n+1}^{\infty} e_k^*(x-y)e_k\|$$

$$= \|\psi_n(t)e_n^*(x-y)e_n + (x-y) - P_n(x-y)\|$$

$$\leq 4\,\|x-y\|, \tag{1.17}$$

where the inequalities $0 \leq \psi_n(t) \leq 1$ and $\|P_n(x-y)\| \leq 2\,\|x-y\|$ have been used.

Since x and u are solutions of (1.13), the uniqueness implies $u(t) = x(t)$ for $t \geq b_m$. Finally, fix $t > 0$ sufficiently small and $n \in N$. Let $m > n$ be such that $x(b_m) \neq 0$ and $b_m < t$. Then $e_n^*(x(t)) = e_n^*(u(t)) = 0$. In other words, $e_n^*(x(t)) = 0$ for all $n \in N$. Therefore

$$x(t) = \sum_{1}^{\infty} e_n^*(x(t))e_n = 0 \text{ for all } t > 0$$

sufficiently small, which is a contradiction.

Situation (II) also yields a contradiction (as below). It follows from the definitions of ϕ_n and ψ_n that, for $n > m$ and $t \in [b_{m+1}, b_m]$, we have $\phi_n(t) = 0$ (which implies $e_n^*(\phi(t,x(t)) = 0)$ and $\psi_n(t) = 1$. Then

$$P(t,x(t)) = \psi_m(t)e_m^*(x(t))e_m + x(t).$$

We suppose that $x(b_m) = x(b_{m+1}) = 0$ for m large enough, and we may assume $x(t) \neq 0$ on $]b_{m+1}, b_m[$. Combining these remarks, we get

$$\frac{d}{dt}(e_n^*(x(t))) = e_n^*(x'(t)) = e_n^*(f(t,x(t)))$$

$$= \begin{cases} \dfrac{e_n^*(x(t))}{\sqrt{\|P(t,x(t))\|}}, & \text{if } P(t,x(t)) \neq 0 \\ 0, & \text{if } P(t,x(t)) = 0 \end{cases} \tag{1.18}$$

for all $n > m$ and $t \in [b_{m+1}, b_m]$. Clearly, this implies

$$e_n^*(x(t)) = 0, \quad \forall n > m, \quad t \in [b_{m+1}, b_m]. \tag{1.19}$$

By (1.19) we see that $x(t) = P_m(t)$ on $[b_{m+1}, b_m]$. We now can easily check that

$$\frac{d}{dt}(e_m^*(x(t))) = \begin{cases} 0, & \text{if } 0 \leqslant t \leqslant a_m \\ \phi_m(t)h((t-b_{m+1})^2/4), & \text{if } a_m \leqslant t \leqslant c_m, \\ \dfrac{\psi_m(t)\, e_m^*(x(t))}{\sqrt{\|P(t,x(t))\|}}, & \text{if } c_m \leqslant t \leqslant b_m. \end{cases}$$

Then it follows (since $b_{m+1} < a_m$) that

$$e_m^*(x(b_{m+1})) = e_m^*(x(a_m)) = 0 < e_m^*(x(c_m)) \leqslant e_m^*(x(b_m)),$$

i.e., $e_m^*(x(b_m)) > 0$, which is a contradiction. The proof is complete.

<u>Remark 1.1</u> Obviously, Theorem 1.1 can be restated as: *The Banach space* X *is finite dimensional if and only if Peano's theorem holds.*

It is very difficult to construct a function F satisfying the conclusion of Theorem 1.1. However, in theory there are many such functions. Precisely, one has

Theorem 1.3 (Pianigiani [1]). *Let* $C = C(R \times X, X)$ *be the space of all continuous functions from* $R \times X$ *into* X *endowed with the supremum norm. Let* $M \subset C$ *be the set of those functions* F *for which the Cauchy problem* $x'(t) = F(t,x)$, $x(t_0) = x_0$ *has no solution. Then* M *is dense in* C *(i.e.,* $\bar{M} = C$).

For the proof of Theorem 1.2, the following equivalent form of Theorem 1.1 is needed.

Theorem 1.4 *Let* X *be any infinite dimensional Banach space. Then for each* $(t_0,x_0) \in R \times X$ *and* $a \in X$ *there exists a continuous function* $F: R \times X \to X$ *such that* $F(t_0,x_0) = a$ *and the Cauchy problem* $x' = F(t,x)$, $x(t_0) = x_0$ *has no solution on any interval* $[t_0, t_0 + d]$ *with* $d > 0$.

Proof Let $\{e_n\}$ and $\{e_n^*\}$ be the sequences given by Day's theorem. We can assume that $e_1 = a/\|a\|$ if $a \neq 0$ (if $a = 0$ then e_1 can be any unit vector).

Set $L_{x_0} = \{z \in X;\ z = x_0 + y,\ y \in L\}$ and let us define $g: R \times L_{x_0} \to L$ as below:

$$g(t,x) = a + \frac{P(t - t_0, x - x_0 - a(t - t_0))}{\sqrt{\|P(t - t_0, x - x_0 - a(t - t_0))\|}} + \phi(t-t_0, x-x_0 -a(t-t_0))$$

if

$$P(t - t_0, x - x_0 - a(t - t_0)) \neq 0,$$

and

$$g(t,x) = a + \phi(t - t_0, x - x_0 - a(t-t_0)),$$

otherwise. As in the proof of Theorem 1.1, one considers a continuous extension F of g to the whole $R \times X$ with values in L. Since $g(t_0,x_0) = a$, then $F(t_0,x_0) = a$.

With such a function F, the conclusion of the theorem holds. Indeed, if we assume that there is a solution x defined in some intervals $[t_0,t_0 + d]$ for $d > 0$, then

$$x(t) = x_0 + \int_{t_0}^{t} F(s,x(s))ds \in x_0 + L.$$

Hence $x(t) \in L_{x_0}$ for $t \in [t_0, t_0 + d]$. Consequently, along the solution x, we have $x'(t) = g(t,x(t))$, $x(t_0) = x_0$. Let us set $s = t - t_0$ and $y = x - x_0 - a(t - t_0)$. Then it follows that y is a solution of (1.6) with f given by (1.5), which is a contradiction.

Proof of Theorem 1.3 Let $f \in C$, $t_0 \in R$ and $x_0 \in X$. Let $F:R \times X \to X$ be a function satisfying the conclusion of Theorem 1.4 with $a = f(t_0,x_0)$ (so $F(t_0,x_0) = f(t_0,x_0)$). Fix $\varepsilon > 0$ and set

$$A = \{(t,x);\ \|F(t,x) - f(t,x)\| \leqslant \varepsilon/2\}$$

$$B = \{(t,x);\ \|F(t,x) - f(t,x)\| \geqslant \varepsilon\}$$

$$C = \{(t,x);\ \varepsilon/2 < \|F(t,x) - f(t,x)\| < \varepsilon\}.$$

Since A and B are disjoint closed sets of $R \times E$, by the well-known Uryson's theorem there exists a continuous real valued function $h:R \times X \to X$ such that $h(y) = 0$ if $y \in B$, $h(y) = 1$ if $y \in A$ and $0 \leqslant h(y) \leqslant 1$ for any $y \in R \times X$. Define $\hat{F}:R \times X \to X$ by

$$\hat{F}(t,x) = h(t,x)F(t,x) + (1 - h(t,x))f(t,x).$$

From $\|F(t_0,x_0) - f(t_0,x_0)\| < \varepsilon/2$, we see that A contains a whole neighbourhood V of (t_0,x_0). Since $\hat{F}(t,x) = F(t,x)$ in V, it follows that the Cauchy problem $x' = \hat{F}(t,x)$; $x(t_0) = x_0$ has no solution (i.e., $\hat{F} \in M$). Finally, inasmuch as

$$\|\hat{F}(t,x) - f(t,x)\| = h(t,x)\,\|F(t,x) - f(t,x)\| \leqslant \|F(t,x) - f(t,x)\|$$

and $R \times X = A \cup B \cup C$, we have $\|\hat{F} - f\|_C \leqslant \varepsilon$, which concludes the proof.

§2. Differential equations associated with continuous and dissipative time-dependent domain operators

2.1 Local existence. A sharp version of the classical method of polygonal lines

In this section we study the existence and uniqueness of solutions (to the right of t_0) of the Cauchy problem

$$u'(t) = A(t)u(t), \quad u(t_0) = x_0 \quad D(A(t_0)), \quad t \geqslant t_0, \tag{2.1}$$

where $A(t): D(A(t)) \to X$ is a nonlinear continuous operator with t-dependent domain $D(A(t)) \subset X$ (for t in the interval $[t_0,b[\subset R$, $t_0 < b \leq +\infty$). To simplify notation, set

$$D(A(t)) \equiv D(t), \quad t \in [t_0,b[\tag{2.2}$$

and

$$\mathcal{D}_T = \{(t,x) \in [t_0,T] \times X, \quad x \in D(t)\}, \quad T \in [t_0,b[. \tag{2.3}$$

The t-dependence of D(t) is restricted to the following conditions:

(A1) For each $t_0 \in]a,b[$ and $x_0 \in D(t_0)$, there exist $r > 0$ and $\bar{T} \in [t_0,b[$ such that $B(x_0,r) \cap D(t)$ is nonempty for each $t \in [t_0,\bar{T}]$.

(A2) The (multivalued) mapping $t \to B(x_0,r) \cap D(t)$ is closed (i.e., if $\{t_n\}_1^\infty$ is a sequence in $[t_0,\bar{T}]$ with $t_n \to t$ and if $x_n \in D(t_n) \cap B(x_0,r)$ is convergent to x, then $x \in D(t) \cap B(x_0,r)$).

The t-dependence of A(t) is restricted to the following conditions:

(A3) The function A (i.e., $(t,x) \to A(t)x$) is continuous from $\mathcal{D}_{\bar{T}}$ into X (i.e., if $(t_n,x_n) \to (t,x)$ with (t_n,x_n), $(t,x) \in \mathcal{D}_{\bar{T}}$, then $A(t_n)x_n \to A(t)x$ as $n \to \infty$).

(A4) $\liminf_{h \downarrow 0} h^{-1}\, d(x + hA(t)x; D(t+h)) = 0$, $\forall t \in [t_0,b[$, $x \in D(t)$. Here $d(x;D)$ denotes the distance from $x \in X$ to the subset $D \subset X$. It is known that we have

$$|d(x;D) - d(y;D)| \leq \|x - y\|, \quad \forall x,y \in X \tag{2.3$'$}$$

(A5) There is a (bounded on bounded subsets) function $L:[t_0,b[\to [0,+\infty[$, such that

$$\langle A(t)x - A(t)y, x-y\rangle_i \leq L(t)\, \|x - y\|^2$$

for all $t \in [t_0,b[$, $x,y \in D(t)$ (i.e., for each $t \in [t_0,b[$, $A(t):D(t) \subset X \to X$ is L(t)-dissipative).

Remark 2.1 In the case D(t) = D-independent of t, the condition (A1) is trivially satisfied if $x_0 \in D$. In this case (A2) means that D is locally closed (in the sense that for each $x_0 \in D$, there is $r > 0$ such that $B(x_0,r) \cap D$ is closed). Here $B(x_0,r)$ is the closed ball of radius r about x_0 (Ch.1, (1.1)).

Moreover, if $A(t)$ is Lipschitz continuous from $D(t)$ into X, i.e.

$$\|A(t)x - A(t)y\| \leq L(t)\ \|x - y\|\ ,\ x,y \in D(t),\ t \in [t_0,b[,$$

then (A5) holds, since

$$\langle A(t)x - A(t)y,\ x - y\rangle_p \leq \|A(t)x - A(t)y\|\ \ \|x - y\|,\ \ p = i \text{ or } s$$

(see the end of Chapter 1). Therefore the classical existence results on the solution to (2.1) are included here.

Finally, Equation (2.1) is actually (1.1) with $A(t) = F(t,\cdot)$ (i.e., $A(t)$ denotes the operator $x \to F(t,x)$). This notation (2.1) is more convenient.

Remark 2.2 If $(t,x) \to A(t)x$ is continuous at (t_0,x_0) then it is locally bounded at this point. This indicates the existence of certain numbers $M > 0$, $\bar{T} \in]t_0,b[$ and $r > 0$, such that

$$\|A(t)x\| \leq M,\ \forall t \in [t_0,\bar{T}[,\ x \in B(x_0;r) \cap D(t). \tag{2.4}$$

In view of (A1), (2.4) makes sense. More precisely, we may choose M, $\bar{T}$ and r (depending on (t_0,x_0)) such that both (A1) and (2.4) are satisfied. □

Definition 2.1 *A function* u *is said to be a solution to* (2.1) *on* $[t_0,T]$ *if* u *is differentiable on* $[t_0,T]$, $u(t) \in D(t)$, $u(t_0) = x_0 \in D(t_0)$ *and* $u'(t) = A(t)u(t)$ *for all* $t \in [t_0,T]$. *If* $t = t_0$, *or* $t = T$, *then* $u'(t)$ *is the appropriate one-sided derivative of* u *at* t. *Since the right-hand side of* (2.1) *is continuous, then* u' *is necessarily continuous on* $[t_0,T]$ (*i.e.*, u *is continuously (strongly) differentiable on* $[t_0,T]$).

Let us now proceed to state the main result of this section.

Theorem 2.1 *Let* $]a,b[$ *be a real interval with* $-\infty \leq a < b \leq +\infty$ *and let* $t_0 \in]a,b[$ *and* $x_0 \in D(t_0)$. *Suppose that conditions* (A1) - (A5) *and* (2.4) *are fulfilled with* $\bar{T} - t_0 = T \leq r/(M+1)$. *Then the Cauchy problem* (2.1) *has a unique solution on* $[t_0,t_0 + T]$. *Moreover, Condition* (A4) *is also necessary for the existence of the solution to* (2.1).

For the proof of this theorem, some preliminary results are needed.

Lemma 2.1 *Assume that conditions* (A1) - (A4) *are satisfied. Let* $t \in [t_0,b[$,

$x \in D(t)$ *and* $\varepsilon > 0$. *Also, let* $r > 0$ *and* $M > 0$ *be such that* $\|A(s)y-A(t)x\| \leq \varepsilon$, $\|A(s)y\| \leq M$ *for all* $|s-t| < r$, $\|y-x\| \leq r$ *with* $y \in D(s)$. *Set* $h_0 = \min\{r, r/M, b-t\}$. *Then*

$$d(x + hA(t)x, D(t+h)) \leq h\varepsilon, \quad \forall h \in]0,h_0[.$$

<u>Proof</u> Let $h \in]0,h_0[$ be fixed and let $\delta \in]0,(r-hM)/h[$. We first prove the existence of a sequence $\{(s_n,y_n)\}$ with $|s_n - t| \leq r$, $y_n \in D(s_n)$, $\|y_n-x\| \leq r$ having the properties

$$s_0 = t, \; y_0 = x, \quad t \leq s_n < s_{n+1} \leq t + h, \quad \lim_{n\to\infty} s_n = t+h \tag{2.5}$$

$$\|y_n + (s_{n+1} - s_n)A(s_n)y_n - y_{n+1}\| \leq (s_{n+1} - s_n)\delta, \quad n \geq 0. \tag{2.6}$$

We shall prove this fact by induction. Choose $s_0 = t$, $y_0 = x$. If $s_n \in [t,t+h[$ and $\|y_n-x\| \leq (s_n-t)(M+\delta)$, define

$$\bar{\sigma}_n = \sup\{\sigma \geq 0; \; s_n + \sigma \leq t+h, \; d(y_n + \sigma A(s_n)y_n; \; D(s_n + \sigma)) \leq \sigma\delta/2\}.$$

Then there is a number $\sigma_n \in]1/2\,\bar{\sigma}_n, \bar{\sigma}_n]$ and an element $y_{n+1} \in D(s_n + \sigma_n)$ such that

$$s_n + \sigma_n \leq t+h, \quad \|y_n + \sigma_n A(s_n)y_n - y_{n+1}\| \leq \delta\sigma_n. \tag{2.7}$$

Set

$$s_{n+1} = s_n + \sigma_n, \; p_n = (y_{n+1} - y_n - \sigma_n A(s_n)y_n)/\sigma_n. \tag{2.8}$$

Clearly, $s_n < s_{n+1}$ provided that $s_n < t+h$. Moreover,

$$y_{n+1} = y_n + \sigma_n(A(s_n)y_n + p_n), \quad \|p_n\| \leq \delta. \tag{2.9}$$

Consequently,

$$\|y_{n+1}-x\| \leq \|y_n-x\| + (s_{n+1}-s_n)(M+\delta) \leq (s_{n+1}-t)(M+\delta).$$

Hence

$$\|y_{n+1}-x\| \leq h(M+\delta) \leq r, \quad |s_{n+1}-t| \leq h \leq r.$$

It remains to show that $\lim_{n\to\infty} s_n = t+h$. Assume by contradiction that $\lim_{n\to\infty} s_n = \bar{s} < t+h$.

By (2.9), $\|y_{n+1}-y_n\| \leqslant (s_{n+1}-s_n)(M+\delta)$ so $\lim_{n\to\infty} y_n = z$ exists. In view of (A2) (with $x_0 = x$ and $t_0 = t$) it follows that $z \in D(\bar{s})$. Then, on the basis of (A4), we can choose a positive number $\eta < t+h-\bar{s}$ such that

$$d(z + \eta A(\bar{s})z;\ D(\bar{s} + \eta)) \leqslant \delta\eta/3. \tag{2.10}$$

On the other hand (by induction hypothesis), $1/2\ \bar{\sigma}_n < \sigma_n \to 0$ as $n \to \infty$, hence $\bar{\sigma}_n < \eta$ for all $n \geqslant n_0$. Let us consider $\eta_n = \bar{s} + \eta - s_n$. We have $\eta_n > \eta > \bar{\sigma}_n$, $\forall n \geqslant n_0$. Then, by the definition of $\bar{\sigma}_n$, one has

$$d(y_n + \eta_n A(s_n)y_n;\ D(s_n + \eta_n)) > \eta_n\ \delta/2. \tag{2.11}$$

Noting that $s_n + \eta_n = \bar{s} + \eta$ (so $\eta_n \to \eta$ and $D(s_n + \eta_n) = D(\bar{s} + \eta)$) and letting $n \to \infty$ in (2.11), one obtains

$$d(z + A(\bar{s})z;\ D(\bar{s} + \eta)) \geqslant \eta\delta/2,$$

which contradicts (2.10). Thus $\lim_{n\to\infty} s_n = t + h$ and therefore $\lim_{n\to\infty} y_n \equiv z \in D(t+h)$.

Let us now proceed to prove the conclusion of the lemma. Obviously,

$$\begin{aligned} d(x + hA(t)x;\ D(t + h)) &\leqslant \|x + hA(t)x - z\| \\ &= \lim_{n\to\infty} \|x + (s_{n+1} - t)A(t)x - y_{n+1}\| \end{aligned} \tag{2.12}$$

Since $s_0 = t$, $y_0 = x$, $(s_{n+1} - t) = \sum_{i=0}^{n} (s_{i+1} - s_i)$ and (by (2.9))

$$y_{n+1} = x + \sum_{i=0}^{n} (s_{i+1}-s_i)(A(s_i)y_i + p_i), \quad \|A(s_i)y_i - A(t)x\| < \varepsilon,$$

we have

$$\|y_{n+1} - x - (s_{n+1} - s)\ A(t)x\| \leqslant (s_{n+1} - s_0)(\varepsilon + \delta). \tag{2.13}$$

Substituting (2.13) in (2.12) we get

$$d(x + hA(t)x;\ D(t+h)) \leqslant h(\varepsilon + \delta)$$

for every $\delta \in]0,(r-hM/h[$. This completes the proof.

Corollary 2.1 *Assume that* (A1) - (A4) *hold. Then for each compact* $K \subset \mathcal{D}_b$ (*see* (2.3))

$$\lim h^{-1} d(x + hA(t)x; D(t+h)) = 0$$

uniformly with respect to $(t,x) \in K$.

Proof One applies Lemma 2.1. Given $\varepsilon > 0$, then M,r (and $b-t$) can be chosen independently of $(t,x) \in K$ (since A is uniformly continuous and bounded on K). It follows that h_o depends only on ε.

Lemma 2.2 *Let* $t \in]a,b[$, $x \in D(t)$ *and* $\varepsilon \in]0,1[$. *Let* $r > 0$ *and* $M > 0$ *be such that* $t + r < b$, $\|A(s)y - A(t)x\| \leqslant \varepsilon/3$ *and* $\|A(s)y\| \leqslant M$ *for all* $|s-t| \leqslant r$, $y \in B(x,r) \cap D(s)$. *Finally, let* $h \in]0,r/(M+1)]$ *and* $\{s_k\}_0^n$ *be a partition of* $[t,t + h]$ (*i.e.*, $t = s_0 < s_1 < \ldots < s_n = t + h$). *Then there exist* $\{y_k \in D(s_k)$, $y_0 = x$, $k = 0,\ldots,n-1\}$ *such that*

$$y_{k+1} = y_k + (s_{k+1}-s_k)(A(s_k)y_k + p_k), \quad \|P_k\| \leqslant \varepsilon. \tag{2.14}$$

Proof Let us assume that $\{y_k, k = 0,1,\ldots,i, i < n-2\}$ satisfy the above properties. We have to define y_{i+1}. Clearly (2.14) yields

$$\|y_k - x\| \leqslant (s_k - t)(M + \varepsilon), \; k = 0,1,\ldots,i. \tag{2.15}$$

Set $\bar{r} = (s_{k+1} - s_i)(M + 1)$ (so $\bar{r} < r$). It is easy to check that if $|s-s_i| \leqslant \bar{r}$ and $\|y - y_i\| \leqslant \bar{r}$, $y \in D(s)$, then $|s-t| \leqslant r$ and $\|y-x\| \leqslant r$. Consequently,

$$\|A(s)y-A(s_i)y_i\| \leqslant \|A(s)y-A(t)x\| + \|A(s_i)y_i-A(t)x \leqslant 2\|\varepsilon/3.$$

According to Lemma 2.1 we have

$$d(y_i + (s_{i+1}-s_i)A(s_i)y_i; D(s_{i+1})) \leqslant 2(s_{i+1}-s_i)\varepsilon/3,$$

which implies the existence of an element, (say) $y_{i+1} \in D(s_{i+1})$, such that

$$\|y_i + (s_{i+1} - s_i)A(s_i)y_i - y_{i+1}\| \leqslant (s_{i+1} - s_i)\varepsilon.$$

With

$$P_i = (s_{i+1} - s_i)^{-1}(y_{i+1} - y_i - (s_{i+1} - s_i)A(s_i)y_i),$$

(2.14) also holds for $k = i$. Therefore we have

$$\|y_{i+1} - x\| \leqslant \|y_i - x\| + (s_{i+1} - s_i)(M + \varepsilon) \leqslant (s_{i+1} - t)(M + \varepsilon),$$

where (2.15) has been used (with $k = i$). □

We now proceed to discuss the problem of the approximate solutions to (2.1). We shall see that some difficulties arise in the construction of approximate solutions u_n. One of these difficulties is the fact that their images (i.e., $u_n(t)$) do not lie in $D(A(t))$ unless $t = t_i^n$. Therefore $A(t)u_n(t)$ does not make sense for $t \neq t_i^n$. In other words, -approximate solutions in the classical sense cannot be expected (in our hypotheses (A1)-(A4)).

So far, the best result in this direction is the following one.

Theorem 2.2 *Suppose that conditions* (A1)-(A4) *hold. Let* $\bar{r} > 0$ *and* $M > 0$ *be such that* $t_0 + \bar{r} < b$ *and* $\|A(t)x\| \leqslant M$ *for all* $t \in [t_0, t_0 + \bar{r}]$ *and* $x \in B(x_0, \bar{r}) \cap D(t)$. *Then for each* $T \in \,]0, \bar{r}/(M+1)]$ *and for every* $\varepsilon \in \,]0,1[$ *there exists an* ε*-approximate solution* u_ε *to* (2.1) *on* $[t_0, t_0 + T]$ *in the following sense: there is a partition*

$$P_\varepsilon = \{t_0 = t_0^\varepsilon < t_1^\varepsilon < \dots < t_{N_\varepsilon - 1} < t_0 + T \leqslant t_{N_\varepsilon}\} \text{ of } [t_0, t_0 + T]$$

having the properties:

(1) $t_{i+1} - t_1 \leqslant \varepsilon,\ 0 \leqslant i \leqslant N_\varepsilon - 1$

where ε *as a superscript for* t_i *is omitted* (*i.e.*, $t_i = t_i^\varepsilon$).

(2) $u_\varepsilon(t_0) = x_0$ *and* $\|u_\varepsilon(t) - u_\varepsilon(s)\| \leqslant |t-s|(M+2\varepsilon/3)$ *for all* $t, s \in [t_0, t_0+T]$.

(3) $u_\varepsilon(t_1) \in B(x_0, \bar{r}) \cap D(t_i),\ 0 \leqslant i \leqslant N_\varepsilon$.

(4) $u_\varepsilon(t) = u_\varepsilon(t_i) + (t-t_i)(A(t_i)u_\varepsilon(t_i) + p_i),\ \|p_i\| \leqslant 2\varepsilon/3,\ t_1 \leqslant t \leqslant t_{i+1}$, *for* $0 \leqslant i \leqslant N_\varepsilon - 1$.

(5) $\|A(t) - A(t_i)u_\varepsilon(t_i)\| \leqslant \varepsilon/3$, *for* $|t - t_i| < (t_{i+1} - t_i)(M+1)$ *and* $\|x - u_\varepsilon(t_i)\| < (t_{i+1} - t_i)(M+1)$ *with* $x \in D(t)$, $i = 0, \dots, N_\varepsilon - 1$.

(6) *Moreover, for any partition* $\{s_k\}_0^n$ *of* $[t_0, t_0+T]$ *with* $\{t_i\}_0^{N_\varepsilon} \subset \{s_k\}_0^n$, *there is* $v_\varepsilon :]t_0, t_0+T] \to X$ *such that* $v_\varepsilon(t_0+) = v_\varepsilon(t_0) = x_0 = u_\varepsilon(t_0)$.

(a) $\|v_\varepsilon(t)-v_\varepsilon(s)\| \leqslant |t-s|(M+\varepsilon)$, $t,s \in]t_i,t_{i+1}]$, $0 \leqslant i \leqslant N_\varepsilon-1$.

(b) $v_\varepsilon(t_i+) = u_\varepsilon(t_i)$, $0 \leqslant i \leqslant N_\varepsilon-1$, *where* $v_\varepsilon(t_i+) = \lim_{h\downarrow 0} v_\varepsilon(t_i+h)$.

(c) $v_\varepsilon(s_k) \in B(x_o,\bar{r}) \cap D(s_k)$; $k = 1,\ldots,n$.

(d) $v_\varepsilon(t) = v_\varepsilon(s_k) + (t - s_k)(A(s_k)v_\varepsilon(s_k) + p_k)$, $\|p_k\| \leqslant \varepsilon$,

$$t \in]s_k,s_{k+1}],\ 0 \leqslant k \leqslant n-1.$$

(e) $\|A(s_{k+1})v_\varepsilon(s_{k+1}) - A(s_k)v_\varepsilon(s_k+)\| \leqslant 2\varepsilon/3$, $0 \leqslant k \leqslant n-1$.

(f) $\|v_\varepsilon(t_{i+1}) - v_\varepsilon(t_{i+1}+)\| = \|v_\varepsilon(t_{i+1}) - u_\varepsilon(t_{i+1})\| \leqslant 2(t_{i+1}-t_i)\varepsilon$,

$$i = 0,\ldots,N_\varepsilon-1.$$

Proof Fix $T \in]0,\bar{r}/(M+1)]$ and $\varepsilon \in]0,1[$. Thus $t_o + T < t_o + \bar{r} < b$. Set $t_o^\varepsilon = t_o$, $x_o^\varepsilon = x_o$ and inductively define t_{i+1}^ε and $x_{i+1}^\varepsilon \in B(x_o,\bar{r}) \cap D(t_{i+1}^\varepsilon)$ as follows: if $t_i^\varepsilon \in [t_o,t_o + T[$ and $x_i^\varepsilon \in B(x_o,\bar{r}) \cap D(t_i)$ satisfies $\|x_i - x_o\| \leqslant (t_i - t_o)(M + 2\varepsilon/3)$, define

$$r_i = \sup\{r;\ r \in]0,\varepsilon[,\ \|A(t)x - A(t_i)x_i\| \leqslant \varepsilon/3, \text{ for all}$$

$$|t - t_i| \leqslant r,\ x \in B(x_i,r) \cap D(t)\},\ t_i^\varepsilon = t_i,\ x_i^\varepsilon = x_i.$$

Set $h_i = r_i/(M+1)$ and $t_{i+1} = \min\{t_i + h_i, t_o + T\}$. If $t_i = t_o + T$, define $x_{i+1} = x_i$ and if $t_i < t_o + T$, let x_{i+1} be an element of $D(t_{i+1})$ with the property

$$\|x_i + (t_{i+1} - t_i)A(t_i)x_i - x_{i+1}\| \leqslant 2(t_{i+1} - t_i)\varepsilon/3. \tag{2.16}$$

The existence of such $x_{i+1} \in D(t_{i+1})$ is ensured by Lemma 2.1. Clearly, with p_i defined similarly to (2.14), we have

$$x_{i+1} = x_i + (t_{i+1} - t_i)(A(t_i)x_i + p_i),\ \|p_i\| \leqslant 2\varepsilon/3 \tag{2.17}$$

and therefore

$$\|x_{i+1} - x_o\| \leqslant \|x_i - x_o\| + (t_{i+1} - t_i)(M + 2\varepsilon/3) \leqslant (t_{i+1}-t_o)(M + 2\varepsilon/3).$$

Next we show that there is a positive integer $N_\varepsilon = N$ such that $t_{N-1} < t_o+T \leqslant t_N$. To this end, assume by contradiction that $t_i < t_o + T$ for all $i \geqslant 0$. Then $\lim_{n\to\infty} (t_i,x_i) = (\bar{t},\bar{x})$ exists and $\bar{x} \in D(\bar{t})$. Since the set $\{(t_i,x_i); i = 0,1,\ldots\}$ is compact, there is a number $r \in]0,\varepsilon]$ such that $\|A(t)x - A(t_i)x_i\| \leqslant \varepsilon/3$

for all $|t - t_i| \leq r$, $x \in B(x_i,r) \cap D(t)$ and for all $i \geq 0$. Then the definition of r_i yields $r_i = (M+1)h_i \geq r$ for all $i \geq 0$. On the other hand, $\lim_{i\to\infty} r_i = \lim_{i\to\infty} h_i = 0$. This contradiction shows the existence of N_ε with the required property. Finally, the function $u_\varepsilon:[t_0,t_{N_\varepsilon}] \to X$ defined by

$$u_\varepsilon(t) = x_i + (t - t_i)(A(t_i)x_i + p_i),\ t_i \leq t \leq t_{i+1} \tag{2.18}$$

with $\|p_i\| \leq 2/3$ (see (2.17)), $i = 0,1,\ldots,N_\varepsilon-1$, satisfies Properties (1)-(5) (since $u_\varepsilon(t_i) = x_i$). To prove the last part of the theorem, let $\{s_k\}_0^n$ be a partition of $[t_0,t_0 + T]$, i.e., $t_0 = s_0 < s_1 < \ldots < s_{n-1} < t_0 + T \leq s_n$, such that $\{t_i\}_0^N \subset \{s_k\}_0^n$. It is sufficient to construct a function $v_\varepsilon:]t_i,t_{i+1}] \to X$ satisfying Properties (6) for each $i = 0,\ldots,N_\varepsilon-1$, with $]t_0,t_0 + T]$ replaced by $]t_i,t_{i+1}]$. With this remark, we may assume (to simplify notation) that $\{s_k\}_0^n$ is a partition of $[t_i,t_{i+1}]$, that is, $t_i \equiv s_0 < s_1 < \ldots < s_{n-1} < s_n \equiv t_{i+1}$.

Let us observe that in view of Property (5) of u_ε, the conditions of Lemma 2.2 hold with (t,x), (s,y) and r replaced by (t_i,x_i), (t,x) and $\bar{r}$ respectively. Consequently, there is a sequence $\{y_k\}_0^n$ such that

$$y_0 = x_i,\ y_k \in D(s_k),\ \|y_k-x_i\| \leq (s_k-t_i)(M+\varepsilon),\ 0 \leq k \leq n \tag{2.19}$$

$$y_{k+1} = y_k + (s_{k+1}-s_k)(A(s_k)y_k + p_k),\ \|p_k\| \leq \varepsilon,\ 0 \leq k \leq n-1 \tag{2.20}$$

$$\|A(s_k)y_k - A(t_i)x_i\| \leq \varepsilon/3,\ 0 \leq k \leq n. \tag{2.21}$$

We have seen that $\|x_i - x_0\| \leq (t_i - t_0)(M + \varepsilon)$, hence

$$\|y_k-x_0\| \leq \|y_k-x_i\| + \|x_i-x_0\| \leq (s_k-t_0)(M+1) \leq \bar{r},\ 0 \leq k \leq n.$$

Define the function $v_\varepsilon:]t_i,t_{i+1}] \to X$ as below:

$$v_\varepsilon(t) = y_k + (t-s_k)(A(s_k)y_k+p_k),\quad t \in]s_k,s_{k+1}],\quad 0 \leq k \leq n-1. \tag{2.22}$$

Therefore

$$v_\varepsilon(t_i+) = v_\varepsilon(s_0+) = y_0 = x_i = u_\varepsilon(t_i),\ v_\varepsilon(s_k) = y_k,$$

$$\|v_\varepsilon(t)-x_0\| \leq \ v_\varepsilon(t)-y_k\| + \|y_k-x_0\| \leq (t-t_0)(M+1) \leq \bar{r}, \tag{2.23}$$

for all $t \in [s_k,s_{k+1}]$.

Properties (a) - (d) are clearly satisfied. Furthermore, if $s_k \in]t_i,t_{i+1}]$ then $v_\varepsilon(s_k+) = v_\varepsilon(s_k) = y_k$ and, by (2.21), we have

$$\|A(s_{k+1})y_{k+1} - A(s_k)v_\varepsilon(s_k+)\| \leq \|A(s_{k+1})y_{k+1} - A(t_i)x_i\|$$

$$+ \|A(t_i)x_i - A(s_k)y_k\| \leq 2\varepsilon/3,$$

that is (6)(e) is also satisfied. It remains to prove (6)(f). To this end, let us observe that (2.22) and (2.18) yield $v'_\varepsilon(t) = A(s_k)y_k + p_k$, $u'_\varepsilon(t) = A(t_i)x_i + p_i$, $t \in]s_k,s_{k+1}[$, where $\|p_i\| \leq 2\varepsilon/3$. Inasmuch as u_ε and v_ε are (Lipschitz) continuous on $[t_i,t_{i+1}]$ and $]t_i,t_{i+1}]$ respectively, one has (in view of (2.21))

$$\|v_\varepsilon(t_{i+1}) - u_\varepsilon(t_{i+1})\| \leq \int_{t_i}^{t_{i+1}} \|v'_\varepsilon(s) - u'_\varepsilon(s)\| \, ds$$

$$(t_{i+1} - t_i)(\varepsilon/3 + \varepsilon + 2\varepsilon/3) = 2(t_{i+1} - t_i)\varepsilon$$

The proof is complete.

Proposition 2.1 *In addition to the hypothesis of Theorem 2.2 assume that for each* $t \in [t_0,t_0 + T]$, *there exists* $\lim_{\varepsilon \downarrow 0} u_\varepsilon(t) = u(t)$. *Then* u *is a solution to* (2.1) *on* $[t_0,t_0 + T]$.

Proof It follows from the property

$$\|u_\varepsilon(t) - u_\varepsilon(s)\| \leq |t-s|(M + 2\varepsilon/3), \quad t,s \in [t_0,t_0 + T]$$

that $\|u(t) - u(s)\| \leq M|t-s|$. Define the step functions a_ε by

$$a_\varepsilon(s) = t_i, \text{ for } s \in [t_i,t_{i+1}[, \ i = 0,\dots,N_\varepsilon-1$$

and $a_\varepsilon(t_0 + T) = t_{N_\varepsilon}$, if $t_0 + T = t_{N_\varepsilon}$. Then we have $u_\varepsilon(a_\varepsilon(s)) = u_\varepsilon(t_i)$, for $s \in [t_i,t_{i+1}[$, hence $u_\varepsilon(a_\varepsilon(s)) \in B(s_0,\bar{r}) \cap D(a_\varepsilon(s))$, $s \in [t_0,t_0 + T]$. Clearly, $|a_\varepsilon(s) - s| \leq \varepsilon$ on $[t_0,t_0 + T]$ and $u_\varepsilon(a_\varepsilon(s)) \to u(s)$ as $n \to \infty$ uniformly on $[t_0,t_0 + T]$, which shows that $u(s) \in B(x_0,\bar{r}) \cap D(s)$. It is also easy to check that u_ε may be rewritten in the following form:

$$u_\varepsilon(t) = x_0 + \int_{t_0}^{t} A(a_\varepsilon(s))u_\varepsilon(a_\varepsilon(s))ds + g_\varepsilon(t), \quad t \in [t_0,t_0 + T] \qquad (2.24)$$

with $g_\varepsilon:[t_0,t_0 + T] \to X$ given by

$$g_\varepsilon(t) = \sum_{j=0}^{i-1} (t_{j+1} - t_j)p_j + (t - t_i)p_i, \; t \in [t_i,t_{i+1}].$$

Since $\|p_i\| \leq 2\varepsilon/3$, it follows that $\|g_\varepsilon(t)\| \leq 2T\varepsilon/3$. Inasmuch as $A(a_\varepsilon(s))u_\varepsilon(a_\varepsilon(s)) \to A(s)u(s)$ as $\varepsilon \downarrow 0$, uniformly on $[t_0,t_0 + T]$, we can pass to the limit in (2.24) as $\varepsilon \downarrow 0$. Doing so, one obtains

$$u(t) = x_0 + \int_{t_0}^{t} A(s)u(s)ds, \; t \in [t_0,t_0 + T] \tag{2.25}$$

which concludes the proof.

<u>Proof of Theorem 2.1</u> Let $T = \min \{b,r/(M+1)\}$ and $c = \sup \{|L(t)|; t \in [t_0,t_0+T]\}$, where L is the function appearing in (A5). Let u_n be the $1/n$-approximate solution given by Theorem 2.2, corresponding to $\varepsilon = 1/n$, $n = 1,2,\ldots$. Denote by $\{t_i^n, 0 \leq i \leq N_n\}$ the partition of $[t_0,t_0 + T]$ corresponding to u_n. Let m and n be positive integers and let $S = \{t_i^n,t_j^m, 0 \leq i \leq N_n, 0 \leq j \leq N_m\}$. Denote by $\{s_k, k = 0,1,\ldots,\bar{N}\}$ the "minimal refinement" of the partitions $\{t_i^n\}$ and $\{t_j^m\}$ inductively defined by

$$s_0 = t_0 \text{ and } s_{k+1} = \min \{s \in S; s > s_k, k = 0,1,\ldots,\bar{N} \leq N_n + N_m\}$$

For each $t \in]t_i^n, t_{i+1}^n] \cap]t_j^m,t_{j+1}^m] \cap]s_k,s_{k+1}]$ we have

$$\begin{aligned}\|u_n(t)-u_m(t)\| &\leq \|u_n(t)-u_n(t_i^n)\| + \|u_n(t_i^n)-v_n(s_{k+1})\| \\ &\quad + \|v_n(s_{k+1})-v_m(s_{k+1})\| + \|v_m(s_{k+1})-u_m(t_j^m)\| + \|u_m(t_j^m)-u_m(t)\| \\ &\leq \|v_n(s_{k+1})-v_m(s_{k+1})\| + 2(M+1)(1/n+1/m).\end{aligned} \tag{2.26}$$

Here we have used the properties of u_n and v_n given by Theorem 2.2. For example,

$$\|u_n(t)-u_n(t_i)\| \leq (t-t_i)(M+2n/3) \leq (t_{i+1}-t_i)(M+1) < (M+1)/n$$

and

$$\|u_n(t_i^n)-v_n(s_{k+1})\| \leq |s_{k+1}-t_i^n|(M + \tfrac{1}{n}) \leq (M+1)/n$$

so (2.26) is now clear. In fact we may assume that

$$t_i^n < t_j^m \leq s_k < t < s_{k+1} = t_{i+1}^m < t_{j+1}^m \quad (\text{or } s_{k+1} = t_{j+1}^m \leq t_{i+1}^n).$$

Thus we need an estimate for $\|v_n(s_{k+1})-v_m(s_{k+1})\|$. For this fact we shall use the dissipativity of A(t) (i.e., Hypothesis (A5)). The reader is urged to keep in mind Inequality (3.4) (in Chapter 1), which is equivalent to ω-dissipativity. $L(s_{k+1})$-dissipativity of A and Theorem 2.2 (part (6)) yield

$$\begin{aligned}
&(1-(s_{k+1}-s_k)c)\,\|v_n(s_{k+1})-v_m(s_{k+1})\| \\
&= (1-(s_{k+1}-s_k)L(s_{k+1}))\,\|v_n(s_{k+1})-v_m(s_{k+1})\| \\
&\leq \|v_n(s_{k+1})-v_m(s_{k+1})-(s_{k+1}-s_k)[A(s_{k+1})v_n(s_{k+1})-A(s_{k+1})v_m(s_{k+1})]\| \\
&\leq \|v_n(s_k+)-v_m(s_k+)\| + (s_{k+1}-s_k)\,[\,\|A(s_{k+1})v_n(s_{k+1})-A(s_k)v_n(s_k+)\| \\
&+ \|p_k^n\| + \|A(s_{k+1})v_m(s_{k+1}) - A(s_k)v_m(s_k+)\| + \|p_k^m\|\,] \\
&\leq \|v_n(s_k+)-v_m(s_k+)\| + 2(s_{k+1}-s_k)(1/n+1/m) \qquad (2.27)
\end{aligned}$$

where $\|p_k^n\| \leq 1/n$, $\|p_k^m\| \leq 1/m$ and

$$v_j(s_{k+1}) = v_j(s_k+) + (s_{k+1}-s_k)(A(s_k)v_j(s_k+) + p_k^j), \quad j \in \{m,n\}.$$

Since $(s_{k+1}-s_k)c \leq c/n \leq 1/2$ and $(1-s)^{-1} \leq \exp 2s$ for all $s \in [0,1/2]$, (2.27) yields

$$\begin{aligned}
\|v_n(s_{k+1})-v_m(s_{k+1})\| &\leq (\,\|v_n(s_k+)-v_m(s_k+)\| + 2(s_{k+1}-s_k) \\
&\qquad (1/n+1/m))\exp(2(s_{k+1}-s_k)c) \\
&\leq [\|v_n(s_k)-v_m(s_k)\| + \|v_n(s_k)-v_n(s_k+)\| + \|v_m(s_k)-v_m(s_k+)\| \\
&+ 2(s_{k+1}-s_k)(1/n+1/m)]\exp(2(s_{k+1}-s_k)c). \qquad (2.28)
\end{aligned}$$

Recall that $v_j(s_0) = v_j(s_0+) = u_j(t_0) = x_0$, $j \in \{m,n\}$. Iterating (2.28) and using Theorem 2.2 (part (f)), one obtains

$$\|v_n(s_{k+1})-v_m(s_{k+1})\| \leq \{ \sum_{e=0}^{k} (\| v_n(s_e)-v_n(s_e+)\| + \|v_m(s_e)-v_m(s_e+)\|)$$

$$+ 2(s_{k+1}-t_o)(\frac{1}{n} + \frac{1}{m})\} \exp (2(s_{k+1}-t_o)c)$$

$$\leq 2\{\frac{1}{n} \sum_{i=1}^{N_n} (t_i^n-t_{i-1}^n) + \frac{1}{m} \sum_{j=1}^{N_m} (t_j^m-t_{j-1}^m)+T(\frac{1}{n} + \frac{1}{m})\}\exp(2cT)$$

$$\leq 4T(\frac{1}{n} + \frac{1}{m}) \exp(2cT). \tag{2.29}$$

Finally, substituting (2.29) into (2.26) we get

$$\|u_n(t)-u_m(t)\| \leq [4T \exp(2cT) + 2(M+1)](\frac{1}{n} + \frac{1}{m}), \ t \in [t_o,t_o+T],$$

which shows that $\lim_{n\to\infty} u_n(t) = u(t)$ exists. The fact that u is a solution to (2.1) has already been proved (Proposition 2.1).

Let us proceed to prove *uniqueness*. Denote by $u = u(t;t_o,x_o)$ a solution of (2.1) on $[t_o,t_o+T]$ with $u(t_o) = x_o$ and $T = T_o = T(t_o,x_o) > 0$. We will prove that

$$\|u(t;t_o,x_o) - u(t;t_o,x_1)\| \leq \|x_o-x_1\| \exp (\int_{t_o}^{t} L(s)ds) \tag{2.30}$$

for all $t \in [t_o,t_o + T_o] \cap [t_o,t_o + T_1]$, $x_o,x_1 \in D(t_o)$.

For this purpose set $v(t) = u(t;t_o,x_o) - u(t;t_o,x_1)$ and $T = \min \{T_o,T_1\}$. Then $t \to \|v(t)\|$ is Lipschitz continuous on $[t_o,t_o + T]$ and therefore it is almost everywhere (a.e.) differentiable on this interval. This remark, the identity $u'(t;t_o,x_o) - u'(t;t_o,x_1) = A(t)u(t;t_o,x_o) - A(t)u(t;t_o,x_1)$, $t \in [t_o,t_o + T]$, and Lemma 2.1 in Chapter 1, lead to the inequality

$$\frac{1}{2}\frac{d}{dt}\|v(t)\|^2 \leq \langle A(t)u(t;t_o,x_o) - A(t)u(t;t_o,x_1),v(t)\rangle_i$$

$$\leq L(t) \|v(t)\|^2, \text{ a.e. on } [t_o,t_o + T].$$

Hence $v(t_o) = x_o - x_1$ and

$$\frac{d}{dt}\|v(t)\| \leq L(t) \|v(t)\|, \text{ a.e. on } [t_o,t_o + T],$$

which yields (2.30). Therefore, if $x_0 = x_1$, then $u(t;t_0,x_0) = u(t;t_0,x_1)$ on $[t_0,t_0 + T]$. It remains to prove the necessity of (A4). Let us assume that for each $t_0 \in]a,b[$ and $x_0 \in D(t_0)$, (2.1) admits a solution $u:[t_0,t_0 + T_0] \to X$ with $T_0 = T(t_0,x_0) > 0$. First of all this means that $u(t_0 + h) \in D(t_0+h)$; hence (see also (2.3)')

$$\frac{1}{h} d(x_0 + hA(t_0)x_0; D(t_0+h)) \leq \frac{1}{h} \|x_0 + hA(t_0)x_0 - u(t_0+h)\|$$

$$= \left\| \frac{u(t_0+h)-u(t_0)}{h} - A(t_0)x_0 \right\|, \quad h > 0.$$

This implies even

$$\lim_{h\downarrow 0} \frac{1}{h} d(x_0 + hA(t_0)x_0; D(t_0+h)) = 0$$

since

$$\lim_{h\downarrow 0} h^{-1}(u(t_0+h) - u(t_0)) = A(t_0)x_0.$$

Remark 2.3 It now follows that under Hypotheses (A1), (A2), (A3) and (A5), Condition (A4) is equivalent to (A4)' below:

$$\lim_{h\downarrow 0} h^{-1}d(x + hA(t)x; D(t+h)) = 0, \quad \forall t \in]a,b[, \ x \in D(t). \qquad \text{(A4)'}$$

Remark 2.4 Without Hypothesis (A5) the conclusion of Theorem 2.1 may not be true. More precisely, Hypotheses (A1)-(A4) do not ensure the convergence of the 1/n-approximate solutions u_n given by Theorem 2.2. Indeed, if, for each t, $D(t) = X$, then (A1), (A2) and (A4) are trivially satisfied while (A3) becomes the usual continuity of the right-hand side of (2.1). However, Godunov's theorem shows that $u' = A(t)u$, $u(t_0) = x_0$ may not have solution (if X is infinite dimensional). The convergence of the approximate solutions is also guaranteed by the compactness assumptions on $(t,x) \to A(t)x$.

Let us discuss some aspects of the continuation of the local solution u to (2.1). Recall that a solution $v:[t_0,t_0 + T_1] \to X$ of (2.1), with $T_1 > T$, is said to be a continuation to the right of the solution $u:[t_0,t_0+T] \to X$ of (2.1), if $v(t) = u(t)$ for all $t \in [t_0,t_0 + T]$. The solution v is said to be noncontinuable if it has no proper continuation. As in the classical theory of ordinary differential equations (in which A(t) is defined on open subsets) one has:

Proposition 2.2 *Assume that, for each* $t_0 \in]a,b[$ *and* $x_0 \in D(t_0)$, (2.1) *has a (local) solution* $u:D(u) \subset]a,b[\to X$. *Then each solution* $u:D(u) \to X$ *of* (2.1) *has a noncontinuable continuation.*

Proof Denote by $\mathcal{F}_u$ the set of all solutions to (2.1) that are continuations to u. Define the partial ordering $\preccurlyeq$ on $\mathcal{F}_u$ in the usual sense, i.e., if $u_1,u_2 \in \mathcal{F}_u$ then $u_1 \preccurlyeq u_2$ iff u_2 is a continuation of u_1. It is easy to prove that every chain $\{u_\alpha\ ;\ \alpha \in \Gamma\}$ in $\mathcal{F}_u$ has an upper bound in $\mathcal{F}_u$. Indeed, set $D = \bigcup_{\alpha\in\Gamma} D(u_\alpha)$. Clearly $D(u) \subset D \subset]a,b[$. Define $v:D \to X$ by $v(t) = u_\alpha(t)$ for $t \in D(u_\alpha)$. Since the chain is totally ordered, one observes that v is well defined (i.e., $u_{\alpha_1}(t) = u_{\alpha_2}(t)$ whenever $t \in D(u_{\alpha_1}) \cap D(u_{\alpha_2})$). Moreover, v is a solution to (2.1) (a continuation of u). By Zorn's lemma, $\mathcal{F}_u$ has a maximal element v*. Such a v* is a noncontinuable continuation of u.

Remark 2.5 It is clear that in the case of the local solution $u:[t_0,t_0+T] \to X$ given by Theorem 2.1, a noncontinuable continuation (to the right) v* is defined on an interval $[t_0,t_0+T_1[$ (with $t_0+T < t_0 + T_1 \leqslant b$) which is open to the right.

2.2 Global existence by means of connected components and relatively open-closed subsets

A classical result on the behaviour of a noncontinuable solution to (2.1) is given by

Proposition 2.3 *Assume in addition to Hypotheses* (A1)-(A4), *that* $t \to D(t)$ *is closed on* $[t_0,b[$, A *maps bounded sets into bounded sets and* (2.1) *has a local solution (to the right) for each* $t_0 \in]a,b[$ *and* $x_0 \in D(t_0)$. *If* $v:[t_0,t_{max}[\to X$, $t_0 < t_{max} \leqslant b$, *is a noncontinuable solution of* (2.1), *then either*

(i) $t_{max} = b$, *or*

(ii) (*if* $t_{max} < b$) $\lim_{t\uparrow t_{max}} \|v(t)\| = +\infty$.

Proof If (i) holds we have nothing to prove. Therefore suppose that $t_{max} \equiv c < b$ and prove (ii). We note that if $c < b$, then $v(t)$ is not bounded on $[t_0,c[$. Indeed, if this were not the case there would be $K > 0$ such that $\|v(t)\| \leq K$, $\forall t \in [t_0,c[$. This would imply $\|A(t)v(t)\| \leq K_1$, $\forall t \in [t_0,c[$. Hence

$$\|v(t)-v(\tau)\| \leq \int_\tau^t \|A(s)v(s)\|\, ds \leq K_1|t-\tau|,\ t,\ \tau \in [t_0,c[, \tag{2.31}$$

so $\lim_{t\uparrow c} v(t) = v_0$ exists. Moreover, $v_0 \in D(c)$. If one defines $\tilde{v}(t) = v(t)$ if $t \in [t_0,c[$ and $\tilde{v}(c) = v_0$, then it is easy to check that v is a solution to (2.1) on $[t_0,c]$ which contradicts the noncontinuability of v. More precisely, the hypothesis $c < b$ implies $\liminf_{t\uparrow c} \|v(t)\| = +\infty$ (which yields (ii)). To prove it, assume for contradiction that there is a sequence $\{t_k\}_1^\infty \subset [t_0,c[$ with $t_k \to c$ and $\|v(t_k)\| \leq r$, $(r > 0)$, for all $k = 1,2,\ldots$. Set $M = \sup\{\|A(t)x\|,\ t \in [t_0,c],\ x \in D(t),\ \|x\| \leq r+1\}$. Then M is a finite positive number (since A is bounded on bounded sets). Let i be a positive integer such that $c - t_k < M^{-1}$ for all $k \geq i$. It follows that

$$\|v(s)\| \leq r+1 \text{ for all } k \geq i \text{ and } s \in [t_k,c[, \tag{2.32}$$

for, if this were not the case, there would be $k \geq i$ and $s_0 \in]t_k,c[$ such that $\|v(s)\| < r + 1$ on $[t_k,s_0[$ and $\|v(s_0)\| = r + 1$. Then

$$\|v(s_0) - v(t_k)\| \leq \int_{t_k}^{s_0} \|A(s)v(s)\|\, ds \leq (s_0-t_k)M < 1,\quad \forall k \geq i,$$

hence

$$1 + r = \|v(s_0)\| \leq (s_0 - t_k)M + \|v(t_k)\| < 1 + r.$$

This contradiction proves (2.32). On the other hand, (2.32) contradicts the unboundedness of v on $[t_0,t_{max}[$. Consequently, if $t_{max} < b$ then $\liminf_{t\uparrow t_{max}} \|v(t)\| = +\infty$, which completes the proof.

Let us now present some conditions under which the local solution u given by Theorem 2.1 can be extended to $[t_0,b[$, that is, $t_{max} = b$. A classical result is the following:

Theorem 2.3 *In addition to Hypotheses* (A1)-(A5), *suppose that* $t \to D(t)$ *is closed on* $[t_0,b[$ *and* A *maps bounded sets into bounded sets. Then for each* $x_0 \in \cap D(t)$, $t_0 \leq t < b$, (*if there exists such* x_0), *the initial value problem* (2.1) *has a unique global solution* $u:[t_0,b[\to X$ (*with* $u(t) \in D(t)$ *for all* $t \in [t_0,b[$).

Proof Let x_0 be as in the statement of the theorem and let $u:[t_0,t_0+T] \to X$ be the local solution to (2.1) given by Theorem 2.1. Denote also by $u:[t_0,c[\to X$ the corresponding noncontinuable solution. We have to prove that $c = b$. Indeed, let us assume by contradiction that $c < b$. Set $M = \sup\{\|A(t)x_0\|\ ;\ t_0 \leq t \leq c\}$. Inasmuch as $t \to A(t)x_0$ is continuous on $[t_0,c]$, we have $M < +\infty$. Clearly,

$$(u(t) - x_0)' = A(t)u(t) - A(t)x_0 + A(t)x_0,\ t \in [t_0,c[.$$

By a standard argument (see, e.g., the proof of (2.30)), the above identity yields

$$\frac{d}{dt}\|u(t)-x_0\| \leq L\,\|u(t) - x_0\| + M, \text{ a.e. on } [t_0,c[.$$

Integrating both sides over $[t_0,t]$, $t \in [t_0,c[$ and then using Gronwall's inequality, one obtains

$$\|u(t) - x_0\| \leq M(c-t_0)\exp L(t-t_0),\ t \in [t_0,c[.$$

where $L = \sup\{L(t),\ t_0 \leq t \leq c\}$. It follows that u is bounded on $[t_0,c[$, which contradicts the conclusion of Proposition 2.3. Therefore $c = b$.

Remark 2.6 For the continuation of the local solution, the condition $x_0 \in \cap D(t)$, $t_0 \leq t \leq b$, is not necessary.

A fundamental result on the Cauchy problem (2.1) is given by Theorem 2.4 below, due to Kenmochi and Takahashi [2].

Theorem 2.4 *Suppose that Hypotheses* (A1)-(A5) *are fulfilled and that* $t \to D(t)$ *is closed on* $[t_0,b[$. *Let* $C(t_0)$ *be a connected component of* $D(t_0)$ *and* $c \in]t_0,b[$. *If there is an* X*-valued continuous function* g *defined on* $[t_0,c[$ *such that* $g(t_0) \in C(t_0)$ *and* $g(t) \in D(t)$ *for all* $t \in [t_0,c[$, *then for every* $z \in C(t_0)$, *the problem* (2.1) *has a unique solution on* $[t_0,c[$.

In the proof of this theorem, the following lemma is needed.

Lemma 2.3 *Suppose that* (A1) - (A5) *are fulfilled and* $t \to D(t)$ *is closed on* $[t_0,b[$. *Let* $s \in [t_0,b[$ *and* $x \in D(s)$. *If* $C(s)$ *is the connected component of* $D(s)$ *containing* x *and if* (2.1) *has a solution* u *on* $[s,\bar{c}]$ *with* $s < \bar{c} < b$ *and* $u(s) = x$, *then for every* $z \in C(s)$, (2.1) *has a solution* u *on* $[s,\bar{c}]$, *with* $u(s) = z$.

Proof For simplicity of writing, take $s = t_0$. Set $Z = \{z \in C(t_0);$ (2.1) has a solution $u = u(t;t_0,z)$ on $[t_0,\bar{c}]\}$. By hypothesis, $u(t_0) = x \in Z$. We shall prove that $Z = C(t_0)$. For this purpose we show that Z is both closed and relatively open in $C(t_0)$. Therefore let $\{z_n\} \subset Z$ be a sequence with $\lim_{n\to\infty} z_n = z_0$. Clearly $z_0 \in C(t_0)$. Moreover, in view of (2.30) we have

$$\|u(t;t_0,z_n) - u(t;t_0,z_m)\| \leq \|z_n - z_m\| \exp \int_{t_0}^{t} L(s)ds,$$

for all $t \in [t_0,\bar{c}]$ and $m, n = 1,\dots$. It follows that $\lim_{n\to\infty} u(t;t_0,z_n) = v(t)$ exists (uniformly on $[0,\bar{c}]$). Inasmuch as

$$u(t;t_0,z_n) = z_n + \int_{t_0}^{t} A(s)u(s;t_0,z_n)ds, \; t \in [t_0,\bar{c}],$$

it is easy to see that v is the solution to Equation (2.1) (with $v(t_0) = z_0$) on $[t_0,\bar{c}]$. Accordingly, $z_0 \in Z$, hence Z is closed in $C(t_0)$. To prove that Z is relatively open, take an arbitrary $z \in Z$. Since $u = u(t;t_0,z)$ is a continuous function on $[t_0,\bar{c}]$, the set $\{(t,u(t;t_0,z); t \in [t_0,\bar{c}]\}$ is compact in $R \times X$. Consequently, there exist two positive numbers M and r such that $\|A(s)y\| \leq M$ for all $(s,y) \in B_r(t,u(t))$ with $y \in A(s)$ and for each $t \in [t_0,\bar{c}]$. Here $B_r(t,u(t))$ is the closed ball in $R \times X$ of radius r about $(t,u(t))$ and $u(t) = u(t;t_0,z)$. Set $k = \sup \{\int_{t_0}^{t} L(s)ds, t \in [t_0,\bar{c}]\}$ and $r_1 = r \exp(-k)$. Then it is easy to check that $C(t_0) \in B(z,r_1) \subset Z$. Indeed, let $y \in C(t_0) \cap B(z,r_1)$ and let $v(t) = u(t;t_0,y)$ be the noncontinuable solution on $[t_0,d[$ to (2.1). We prove by contradiction that $\bar{c} \leq d$. Therefore, suppose that $d < \bar{c}$. By (2.30) we have

$$\|u(t;t_0,y)-u(t;t_0,z)\| \leq \|y-z\| e^k \leq r_1 e^k = r, \; t \in [t_0,d[,$$

so $(t,v(t)) \in B_r(t,u(t;t_0,z))$, which implies $\|A(t)v(t)\| \leqslant M$, $t \in [t_0,d[$. Since $v'(t) = A(t)v(t)$ on $[t_0,d[$, it follows that $\lim_{t\uparrow d} v(t)$ exists, which contradicts the noncontinuability of v. This completes the proof.

Proof of Theorem 2.4 In view of Theorem 2.1, the only fact we have to prove is that the local solution u to (2.1) can be extended to the whole $[t_0,c[$. For this purpose, assume by contradiction that the domain of the noncontinuable solution u to (2.1) with $u(t_0) = z \in C(t_0)$, is $[t_0,d[$, with $t_0 < d < c$. Since the subset $\{(s,g(s));\ s \in [t_0,d]\}$ of $R \times X$ is compact, there are $r > 0$, $M > 0$ and $T > 0$ such that $T < d - t_0$ and $\|A(t)y\| \leqslant M$, $\forall (t,y) \in B_r(s,g(s))$, with $y \in D(t)$ and for each $s \in [t_0,d]$, where

$$B_r(s,g(s)) = \{(t,y) \in R \times X,\ |t-s| \leqslant r,\ \|y-g(s)\| \leqslant r\}. \tag{2.33}$$

Moreover, we may assume that $(M + 1)T \leqslant r$. It follows from Theorem 2.1 that, for each $s \in [t_0,d]$, there is a unique solution $u = u(t;s,g(s))$ to (2.1) on $[s,s + T]$ (with T independent of $s \in [t_0,d]$) corresponding to the initial condition $u(s) = u(s;s,g(s)) = g(s)$. Moreover,

$$\|u(t;s,g(s)) - g(s)\| \leqslant M(t-s),\ t \in [s,s + T]. \tag{2.34}$$

For each $t \in [t_0,c[$ denote by C(t) the connected component of D(t) containing g(t). To apply Lemma 2.3, we first prove that $g(t_0 + T)$ and $u(t_0+T;t_0,g(t_0))$ belong to the same connected component $C(t_0 + T)$ of $D(t_0 + T)$. For this purpose, set $f(t) = u(t_0 + T;t,g(t))$. Then $f:[t_0,t_0+T] \to D(t_0+T)$, $f(t_0) = u(t_0 + T;t_0,g(t_0))$ and $f(t_0 + T) = g(t_0 + T)$. Therefore it suffices to show that f is continuous on $[t_0,t_0 + T]$.

Indeed, for $t_0 \leqslant s \leqslant t \leqslant t_0 + T$, we have

$$\|f(t)-f(s)\| = \|u(t_0+T;t,g(t)) - u(t_0+T;t,u(t;s,g(s)))\|$$

since the uniqueness implies

$$u(t_0+T;s,g(s)) = u(t_0+T;t,u(t;s,g(s))). \tag{2.34'}$$

Consequently, by (2.30) and (2.34),

$$\|f(t)-f(s)\| \leqslant \|g(t)-u(t;s,g(s))\| \exp \int_t^{t_0+T} L(\tau)d\tau$$

$$\leqslant (\|g(t)-g(s)\| + M(t-s) \exp \int_t^{t_0+T} L(\tau)d\tau. \qquad (2.35)$$

By Theorem 2.1, there is a unique solution $u = u(t;t_0 + T,g(t_0+T))$ on $[t_0+T; t_0 + 2T]$. Hence, by Lemma 2.3 with $s = t_0 + T$ and $z = u(t_0+T;t_0,g(t_0))$, the solution $u = u(t,t_0,g(t_0))$ can be extended to $[t_0,t_0 + 2T]$. Applying once again Lemma 2.3 (with $s = t_0$, $x = g(t_0) \in C(t_0)$ and $\bar{c} = t_0 + 2T$), it follows that, for each $z \in C(t_0)$, the solution $u = u(t,t_0,z)$ is defined at least on $[t_0,t_0 + 2T]$. Repeating these arguments we see that the noncontinuable solution to (2.1), $u = u(t,t_0,z)$ on $[t_0,d[$, is defined on $[t_0,t_0+nT]$, $n = 1,2,\ldots$. Since $d < +\infty$, this is a contradiction, so $d = c$.

§3. Generation of nonlinear semigroups on closed subsets (Martin's theorem). Fixed point theorems via the Browder-Kirk theorem

Let D be a subset of X. Recall that D is said to be locally closed if, for each $x \in D$, there is $r > 0$ such that $D \cap B(x,r)$ is closed. In this section we study the existence and the uniqueness of the solution (to the right of t_0) to the initial value problem

$$u' = A(t)u, \quad u(t_0) = x_0 \in D, \; t \geqslant t_0, \qquad (3.1)$$

where $A(t):D \to X$ for each $t \in]a,b[$, $-\infty \leqslant a \quad b \leqslant +\infty$.

A basic hypothesis on A(t) is the following one. There is a function $L:]a,b[\to R$, bounded on bounded sets, such that, for each $t \in]a,b[$, A(t) is L(t)-dissipative, i.e.,

$$\langle A(t)x - A(t)y, \; x - y\rangle_i \leqslant L(t) \; \|x - y\|^2, \quad \forall x,y \in D. \qquad (3.2)$$

The tangential condition (A4) becomes

$$\liminf_{h\downarrow 0} h^{-1}d(x + hA(t)x;D) = 0, \quad \forall t \in]a,b[, \; x \in D. \qquad (3.3)$$

Since we have supposed that $D(A(t)) = D$, $\forall t \in]a,b[$, Theorem 2.1 yields:

Theorem 3.1 *Let* D *be locally closed and let* $(t,x) \to A(t)x$ *be a continuous function from* $]a,b[\times D$ *into* X *satisfying also Condition* (3.2). *Then* (3.3) *is a necessary and sufficient condition for every* $(t_0,x_0) \in]a,b[\times D$ *to have a local solution* u *of* (3.1). *Precisely, given* $t_0 \in]a,b[$ *and* $x_0 \in D$, *let* $M > 0$, $r > 0$ *and* $T > 0$ *be such that* $(M+1)T \leq r$, $t_0 + T < b$ *and* $\|A(t)x\| \leq M$ *for all* $t \in [t_0,t_0 + T]$, $x \in D \cap B(x_0,r)$. *Then* (3.1) *has a unique solution* $u: [t_0,t_0 + T] \to D$.

Concerning the continuation of the local solution to the whole $[t_0,b[$, we have:

Theorem 3.2 *Suppose that* D *is closed, and the continuous function* $A:[t_0,b[\times D \to X$ *satisfies* (3.2) *and* (3.3). *Then the problem* (3.1) *has a unique solution* $u:[t_0,b[\to D$.

Clearly, this theorem is a direct consequence of Theorem 2.4 with c = b and $g(t) = x_0$.

Let us consider the Cauchy problem for the autonomous differential equation

$$u'(t) = Au(t), \quad u(t_0) = x_0 \in D, \quad t_0, t \in [0, +\infty[, \tag{3.4}$$

with D(A) = D and $t \geq t_0$.

Applying Theorem 3.2 with A(t) = A-independent of t, we get:

Theorem 3.3 (Martin) *Suppose that* D *is closed and* $A:D \to X$ *is a continuous and* ω*-dissipative operator, i.e.*

(1) $\langle Ax - Ay, x - y\rangle_i \leq \omega\|x - y\|^2, \quad \forall x,y \in D.$

Then for each $t_0 \in [0,+\infty[$ *and* $x_0 \in D$, *the problem* (3.4) *has a unique solution* $u:[t_0,+\infty[\to D$, *if and only if for each* $x \in D$, Ax *is "tangent" to* D *at* x, *i.e.*

(2) $\liminf_{h\downarrow 0} h^{-1} d(x + hAx;D) = 0, \quad \forall x \in D.$

For each $x_0 \in D$ denote by $u(t) = u(t;0,x_0)$ the solution of (3.4) which corresponds to x_0, that is $u(0) = x_0$. Set $S(t)x_0 = u(t;0,x_0)$. Then for each $t \geq 0$, $S(t):D \to D$ and the following properties hold:

$$S(t)S(s)x = S(t+s)x, \quad \forall x \in D,\ t,s \geq 0, \tag{3.5}$$

$$\lim_{t\to t_0} S(t)x = S(t_0)x,\ S(0)x = x, \quad \forall x \in D,\ t_0 \geq 0. \tag{3.5)'$$

Moreover, by (2.30) we have

$$\|S(t)x - S(t)y\| \leq e^{\omega t} \|x - y\|, \quad \forall x,y \in D,\ t = 0. \tag{3.5)"}$$

Definition 3.1 *The one-parameter family* $\{S(t):D \to D,\ t \geq 0\} = S$ *satisfying* (3.5) *and* (3.5)' *is said to be a semigroup on* D. *If in addition* (3.5)" *is also true, then* S(t) *is said to be a semigroup of type* ω *on* D.

In other words, if $A:D \to X$ satisfies the hypotheses of Theorem 3.3 then A is a "generator" of a semigroup S(t) of type ω on D. In the case in which A is dissipative (i.e., ω = 0) then the semigroup S(t) satisfies (3.5)" with ω = 0 and S(t) is said to be "nonexpansive". At this moment the following definition is needed:

Definition 3.2 *The continuous operator* $A:D \to X$ *is said to be a generator of the differentiable semigroup* S(t) *on* D, *if for each* $x_0 \in D$ *the function* $u(t;0,x_0) = S(t)x_0$ *is a solution on* $[0,+\infty[$ *to the problem* (3.4).

The characterization of continuous generators of semigroups S(t) of type ω on D is given by:

Proposition 3.1 *Suppose that* D *is a closed subset of* X *and* $A;D \to X$ *is continuous. Then the following two conditions are equivalent:*

(1) A *is the generator of the differentiable semigroup* S(t) *on* D *of type* ω.

(2) $\langle Ax-Ay, x-y\rangle_i \leq \omega\|x-y\|^2$ *and* $\liminf_{h \downarrow 0} h^{-1}d(x+hAx;D) = 0$, $\forall x, y \in D$.

Proof In view of Theorem 3.3, the only fact we have to prove is that (1) implies the ω-dissipativity of A. Indeed, by (3.5)" we have

$$\langle x^*, \frac{S(t)x-x}{t} - \frac{S(t)y-y}{t}\rangle \leq \frac{e^{\omega t}-1}{t}\|x-y\|^2, \quad \forall x,y \in D,\ t > 0,\ x^* \in F(x-y).$$

Letting $t \downarrow 0$, we get

$$\langle x^*, Ax - Ay\rangle \leq \omega\|x-y\|^2, \quad \forall x,y \in D,\ x^* \in F(x-y) \tag{3.6}$$

and the proof is complete.

The fundamental result on the generation of nonlinear semigroups (in non-continuous case) is the following one

Theorem (Crandall and Liggett). *Let* $A \subset X \times X$ *be* ω*-dissipative which* *satisfies the range condition*

$$\text{(RC)} \quad R(I - \lambda A) \supset \overline{D(A)}, \tag{3.7}$$

for all sufficiently small positive λ. *Then*

$$\lim_{n\to\infty} (I - \frac{t}{n}A)^{-n}x = S(t)x \in \overline{D(A)}, \quad \forall x \in \overline{D(A)},\ t \geq 0 \tag{3.8}$$

exists uniformly with respect to t *in compact subsets and* S(t) *is a semigroup of type* ω *on* $\overline{D(A)}$. *Moreover, if* A *is closed then Conditions* (i) *and* (ii) *below are equivalent:*

(i) $u:[0,\infty[\to X$ *is a strong solution to* $u' \in Au$, $u(0) = x \in \hat{D}(A)$.

(ii) $u(t) = S(t)x$ *and* $t \to S(t)x$ *is strongly differentiable* a.e. *on* $]0,\infty[$.
Here $\hat{D}(A)$ is the generalized domain of A (Crandall [2]).

In the theory of the generation of nonlinear semigroups the following conditions are the most important: the range condition (3.7) and the tangential conditions below:

$$\lim_{h\downarrow 0} h^{-1}\, d(x + hAx; D) = 0, \quad \forall x \in D, \tag{3.9}$$

$$\lim_{h\downarrow 0} h^{-1}\, d(x; R(I - hA)) = 0, \quad \forall x \in \overline{D(A)}. \tag{3.10}$$

Remark 3.1 In the hypotheses of Theorem 3.3, Conditions (3.3) and (3.9) are equivalent. Indeed, according to this theorem we have

$$h^{-1}d(x+hAx;D) \leq h^{-1}\,\|x+hAx - u(h)\| \leq \|\frac{u(h)-u(0)}{h} - Ax\|, \ \forall x \in D,$$

where u is the solution to (3.4) with $u(0) = x$. Letting $h \downarrow 0$ we get (3.9). With the proof of Proposition 3.1 we see that if $A:D \to X$ is continuous, then the dissipativity of A (i.e., the existence of $x^* \in F(x-y)$ satisfying (3.6)) implies (3.6) for all $x^* \in F(x-y)$.

Lemma 3.1 *Suppose that* $D(A) = D$ *is closed and* $A:D \to X$ *is continuous. Then* (3.9) *implies* (3.10). *If, in addition,* A *is dissipative, then* (3.9) *and* (3.10) *are equivalent.*

Proof We first observe that (3.9) and (3.10) are equivalent to the following, respectively; for each $x \in D$ there exists $g:[0,+\infty[\to X$, such that $g(h) \to 0$ as $h \downarrow 0$ and

$$y_h = x + h(Ax + g(h)) \in D; \tag{3.9)'}$$

for each $x \in D(A)$ and for each $h > 0$ there is $x_h \in D(A)$ such that

$$a(h) = (x_h - hAx_h - x)/h \to 0 \text{ as } h \downarrow 0. \tag{3.10)'}$$

Indeed, in view of (3.9) and (3.10), for each $h > 0$ there exist $y_h, x_h \in D$ with the properties (respectively):

$$\|x + hAx - y_h\| \leq d(x + hAx;D) + h^2,$$

$$\|x - (x_h - hAx_h)\| \leq d(x;R(I-hA)) + h^2.$$

It is now clear that (3.9)' implies (3.10)' with $x_h = y_h$. Conversely, if (3.10)' holds, then the dissipativity of A implies the boundedness of Ax_h, since

$$\|x_h - x\| \leq \|x_h - hAx_h - (x - hAx)\| .$$

Consequently, with $y_h = x_h$, (3.9)' is satisfied.

Obviously, the range condition (3.7) implies (3.10). The converse implication is not true. To this goal we give:

<u>Example 3.1</u> Let $X = R^2$. For $x = (x_1,x_2) \in R^2$, define $Ax = (-x_2,x_1)$ with $D(A) = S(r) = \{x \in R^2, \|x\| = r\}$. Clearly, $A:S(r) \to S(r)$ is continuous and dissipative, since $\langle Ax-Ay,x-y\rangle = 0$. Moreover, (3.9) is verified because $\langle Ax,x\rangle = 0$. See also Lemma 3.2.

However, (3.7) is not true and furthermore we have

$$D(A) \cap R(I - hA) = \phi \ , \ h \in \,]-\infty,+\infty[.$$

Indeed, if $y \in S(r) \cap R(I - hA)$, then there exists $x \in S(r)$ such that $y = x - hAx$, which yields $r^2 = r^2 + h^2r^2$.

<u>Lemma 3.2</u> *On every real Banach space* X, *the following conditions are equivalent:*

(1) $\liminf_{h\downarrow 0} h^{-1}d(x + hy; S(r)) = 0,\ y \in X.$

(2) $\langle x,y\rangle_+ = \lim_{h\downarrow 0} (\|x + hy\| - \|x\|)/h = 0,\quad \|x\| = r.$

(3) $\lim_{h\downarrow 0} h^{-1}d(x + hy; S(r)) = 0,\quad y \in X.$

(*In particular, if* X *is a real Hilbert space* H *of inner product* $\langle\cdot,\cdot\rangle$ *then* (1) *is equivalent to* $\langle x,y\rangle = 0$, *since* $\langle x,y\rangle = \|x\|\,\langle x,y\rangle_+$).

<u>Proof</u> On the basis of (3.9)' we see that (1) is equivalent to the following: there are $h_n \downarrow 0$ and $z_n \in X$ with $z_n \to 0$ as $n \to \infty$, such that $\|x+h_n(y+z_n)\| = r$, therefore $\|x\| = r$. Consequently,

$$\big|\|x + h_n y\| - \|x\|\big| = \big|\|x + h_n y\| - \|x + h_n(y + z_n)\|\big| \leq h_n\,\|z_n\|,$$

which shows that (1) implies (2). To prove that (2) implies (3), choose g(h) with the property

$$x + hy + hg(h) = r(x + hy)/\,\|x + hy\|.$$

Obviously, $\|g(h)\| = \big|\|x + hy\| - \|x\|\big|/h$ and therefore (3) is a consequence of (2). The proof is complete.

We now show how Theorem 3.3 can be applied to the theory of fixed points of nonlinear operators on closed subsets.

<u>Proposition 3.2</u> *Let* D *be a closed subset of* X *and let* $A:D \to X$ *be a continuous operator with the properties:*

(a) $\langle Ax - Ay, x - y\rangle_i \leq \omega\|x - y\|^2,\quad \forall x,y \in D.$

(b) $\lim_{h\downarrow 0} h^{-1}\, d(x + hAx;D) = 0,\quad \forall x \in D.$

If $\omega < 0$, *then there exists* $x_0 \in D$ *such that* $Ax_0 = 0$.

<u>Proof</u> Let x be an arbitrary element of D. On the basis of Theorem 3.3, there exists a unique continuously differentiable function $u:[0,+\infty[\ \to D$ such that

$$u'(t) = Au(t),\ u(0) = x,\ t \geq 0. \tag{3.11}$$

By standard arguments (see the proof of (2.30)), (3.11) yields

$$\frac{d}{dt}\|u(t + h) - u(t)\| \leqslant \omega\|u(t+h) - u(t)\|, \text{ a.e. on } [0,+\infty[. \tag{3.12}$$

Hence

$$\|u(t + h) - u(t)\| \leqslant \|u(h) - x\| e^{\omega t}, \forall t, h > 0. \tag{3.13}$$

Let us write (3.11) in the form

$$(u(t) - x)' = Au(t) - Ax + Ax.$$

Arguing as above, we get

$$\frac{d}{dt}\|u(t) - x\| \leqslant \omega\|u(t) - x\| + \|Ax\|, \text{ a.e. on } [0,+\infty[.$$

Solving this differential inequality, one obtains

$$\|u(t) - x\| \leqslant -\omega^{-1} \|Ax\| (1 - e^{\omega t}), \ \forall t \geqslant 0. \tag{3.14}$$

Combining (3.13) and (3.14) we derive the existence of both $\lim_{t\to\infty} u(t) = x_0$ and $\lim_{t\to\infty} \|u'(t)\| = 0$. Letting $t \to \infty$ in (3.11), the result follows.

Corollary 3.1 *Let* D *be closed and let* $B:D \to X$ *be a continuous operator such that:*

(1) $\lim_{h\downarrow 0} h^{-1} d(x + h(Bx - x);D) = 0, \quad \forall x \in D.$

(2) $\langle Bx - By,x - y\rangle_i \leqslant k \|x - y\|^2, \quad \forall x,y \in D$

for some $k < 1$ (*i.e.*, B *is* k-*dissipative*).

Then B *has a unique fixed point in* D.

Proof Set $A = B - I$. Then A is k-1 dissipative and, by Proposition 3.2, there exists $x_0 \in D$ such that $Bx_0 - x_0 = 0$. □

In particular, if $B:D \to D$ and D is convex then the tangential condition (1) in Corollary, 3.1 is trivially satisfied. Consequently, the following result holds:

Corollary 3.2 *Let* D *be a closed convex subset of* X. *If* $B:D \to D$ *is continuous and* k*-dissipative with* $k < 1$, *then* B *has a unique fixed point in* D.

Recall that $B:D \to D$ is said to be a k-contraction if

$$\|Bx - By\| \leqslant k\,\|x - y\|, \quad k < 1, \quad x,y \in D. \tag{3.15}$$

The well-known Banach principle asserts that any k-contraction B from a closed subset D into itself has a unique fixed point in D. Comparison of the Banach principle with Corollary 3.2 is left to the reader.

Lemma 3.3 *Suppose that* D *is a closed subset of* X *and* $A:D \to X$ *is continuous, dissipative and satisfies the tangential condition* (b) *in Proposition* 3.2. *Let* S *be the nonexpansive semigroup on* D *generated by* A *(in the sense of Definition* 3.2). *Then the function* $t \to \|(AS(t)x\|$ *is nonincreasing on* $[0,+\infty[$, *for each* $x \in D$.

Proof In view of (3.12), with $\omega = 0$ and $u(t) = S(t)x$, we have

$$\frac{d}{dt}\|S(t+h)x - S(t)x\| \leqslant 0, \; h > 0, \text{ a.e. on } [0,+\infty[.$$

Integrating over $[s,t]$, with $0 \leqslant s \leqslant t$, it follows that

$$\|S(t + h)x - S(t)x\| \leqslant \|S(s + h)x - S(s)x\|.$$

Dividing by $h > 0$ and then letting $h \downarrow 0$, one obtains

$$\|\frac{d^+}{dt} S(t)x\| \leqslant \|\frac{d^+}{dt} S(s)x\|, \quad 0 \leqslant s \leqslant t$$

and the proof is complete.

Definition 3.3 *The semigroup* S *on* D *is said to be bounded if for each* $x \in D$ *there exists* $M(x) > 0$ *such that* $\|S(t)x\| \leqslant M(x)$, $\forall t \geqslant 0$. S *has a fixed point* $x_0 \in D$, *if* $S(t)x_0 = x_0$, *for all* $t > 0$.

Using Proposition 3.2 we can easily derive the following:

Theorem 3.4 (Martin). *If* $B:X \to X$ *is continuous and dissipative, then* $R(I - B) = X$.

Proof Let y be arbitrary in X. Set $A = B - I + y$. According to Proposition

3.2 (with X in place of D and $\omega = -1$), there exists $x \in X$ such that $Bx - x + y = 0$. □

The following result is also useful.

Proposition 3.3 *Suppose that* $D \quad X$ *is closed and* $A:D \to X$ *is continuous, dissipative and satisfies the tangential condition* (3.9). *If in addition* $\|Ax\| \to \infty$ *as* $\|x\| \to \infty$, *then all the solutions to* $u' = Au$ *are bounded on* $[0,+\infty[$ *(i.e., the semigroup* S *generated by* A *is bounded).*

Proof If we assume that S is not bounded then there exist $x \in D$ and $t_n > 0$, such that $\|S(t_n)x\| \geq n$, $n = 1,2,\ldots$. By one of the hypotheses it follows that $\|AS(t_n)x\| \to \infty$ as $n \to \infty$, which contradicts the conclusion of Lemma 3.3.

In what follows, the well-known result below is needed.

Theorem (Browder-Kirk). *Let* D *be a closed convex bounded subset of the uniformly convex space* X. *If* $B:D \to D$ *is a nonexpansive operator (i.e.,* $\|Bx - By\| \leq \|x - y\|$, $\forall x,y \in D$), *then* B *has a fixed point.*

Some information on the set of all fixed points of B is given by:

Lemma 3.4 *If* D *is a convex subset of the strictly convex space* X *and* $B:D \to X$ *is nonexpansive, then* $F(B) \equiv \{x \in D;\ x = Bx\}$ *is convex and closed.*

Proof If F(B) is empty or consists of a single element, then we have nothing to prove. Consequently, suppose that $x_1,x_2 \in F(B)$ with $x_1 \neq x_2$. Set $x_\lambda = (1-\lambda)x_1 + \lambda x_2$, $0 < \lambda < 1$. We prove that $x_\lambda \in F(B)$. With $r = \|x_1-x_2\|$ we have

$$\|Bx_\lambda-x_1\| = \|Bx_\lambda-Bx_1\| \leq \|x_\lambda-x_1\| = \lambda r,\ \|Bx_\lambda-x_2\| \leq (1-\lambda)r.$$

Therefore

$$\|x_1-x_2\| \leq \|x_1-Bx_\lambda\| + \|Bx_\lambda-x_2\| \leq \|x_1-x_2\|,$$

that is,

$$\|Bx_\lambda-x_1\| + \|Bx_\lambda-x_2\| = \|x_1-x_2\| \equiv r. \tag{3.16}$$

It follows that

$$||Bx_\lambda - x_1|| = \lambda r \equiv r_1, \quad ||Bx_\lambda - x_2|| = (1-\lambda)r = r_2,$$

which means that $Bx_\lambda \in S(x_1,r_1) \cap S(x_2,r_2)$. According to Proposition 1.1 (Chapter 1), $Bx_\lambda = \lambda_0 x_1 + (1-\lambda_0)x_2$ where $\lambda_0 = r_2/(r_1+r_2) = 1-\lambda$. Consequently, $Bx_\lambda = x_\lambda$. Hence $F(B)$ is convex.

Now we can easily prove:

Theorem 3.5 *Let* D *be a bounded closed convex subset of the uniformly convex space* X *and let* $\mathcal{F}$ *be a collection of nonexpansive operators* $T:D \to D$. *If we assume in addition that* $T_1T_2 = T_2T_1$ *for all* $T_1, T_2 \in \mathcal{F}$, *then there exists* $x_0 \in D$ *such that* $Tx_0 = x_0$, *for each* $T \in \mathcal{F}$.

Proof Take an arbitrary $T_1 \in \mathcal{F}$. By Lemma 3.4, $F(T_1)$ is convex and closed. Clearly, $T_2:F(T_1) \to F(T_1)$, for every $T_2 \in \mathcal{F}$. On the basis of Browder-Kirk's theorem there exists $x \in F(T_1)$ such that $T_2x = x$. In other words $F(T_1) \cap F(T_2) \neq \emptyset$. It follows that $\{F(T), T \in \mathcal{F}\}$ has the finite intersection property. Since D is weakly compact and each F(T) is weakly closed, there is $x_0 \in \bigcap_{T\in\mathcal{F}} F(T)$. The proof is complete. We are now prepared to prove:

Theorem 3.6 *Let* D *be a closed convex subset of the uniformly convex space* X *and let* S *be a nonexpansive semigroup on* D. *Then the following two properties are equivalent:*

(i) S *is bounded (in the sense of Definition* 3.3*).*

(ii) S *has a fixed point* $x_0 > D$.

Proof The implication (ii) ⇒ (i) is obvious. Conversely, suppose that (i) holds. Accordingly, for $y \in D$ there is $M(y) > 0$ such that

$$S(t)y < M(y), \quad \forall t \in 0.$$

Set

$$V(y) = \{z \in D; ||z-y|| < M(y) + ||y||\}, \quad F_s = \bigcap_{t \geq s} V(S(t)y).$$

Clearly $y \in F_s$, so F_s is nonempty. Finally set $D_0 = \bigcup_{s \geq 0} F_s$. It is easy to check that $S(t):D_0 \to D_0$, $\forall t \geqslant 0$. Indeed, if $x \in F_s$, then $S(t)x \in F_{s+t}$ for all $t \geqslant 0$. It is also an elementary fact that D_0 is convex and bounded. Since $S(t):\bar{D}_0 \to \bar{D}_0$, it follows from Theorem 3.5 that there exists $x_0 \in \bar{D}_0 \subset D$ such that $S(t)x_0 = x_0$, for all $t \geqslant 0$. □

In what follows we point out some consequences of Theorem 3.6.

Proposition 3.4 *Suppose that* X *is uniformly convex,* D *is a closed convex subset of* X *and* $B:D \to X$ *is a continuous operator such that for each* $x \in D$, $Bx - x$ *is tangent to* D *at* x. *In addition,*

(1) $\langle Bx - By, x-y \rangle_i \leqslant \|x-y\|^2$, $\forall x,y \in D$.

(2) $B - I$ *is unbounded on unbounded subsets of* D.

Then B *has a fixed point.*

Proof By Proposition 3.3 the semigroup S generated by the dissipative operator B-I is bounded. In view of Theorem 3.6, there exists $x_0 \in D$ such that $S(t)x_0 = x_0$ for all $t \geqslant 0$. Since $u(t) = S(t)x_0$ is a solution to $u' = Bu-u$, it follows that $Bx_0 = x_0$.

Corollary 3.3 *Let* D *be a closed convex bounded subset of the uniformly convex space* X. *If* $B:D \to D$ *is continuous and* B-I *is dissipative, then* B *has a fixed point.*

Proof In this case the semigroup S generated by B-I is bounded, so it has a fixed point $x_0 \in D$. □

Remark 3.2 Let D be a closed convex bounded subset of the uniformly convex space X. With the proof of Proposition 3.4, we have actually proved that if $A:D \to X$ is continuous and dissipative and satisfies (3.9) then there exists $x_0 \in D$ such that $Ax_0 = 0 \in X$. If in addition $A:D \to D$, then of course $0 \in D$.

Remark 3.3 If, in addition to the hypotheses of Theorem 3.1, we suppose that $D \cap B(x_0,r)$ is convex, then the proof of this theorem can essentially be simplified. Indeed, let us observe that the ε-approximate solution (to the problem (3.1)) given by(2.18) can be written in the form (with $\varepsilon = 1/n$)

$$(t^n_{i+1}-t^n_i)^{-1}u_n(t) = (t-t^n_i)x^n_{i+1} + (t^n_{i+1} - t)x^n_i,\ t^n_i \leq t \leq t^n_{i+1} \qquad (3.17)$$

for $i = 0,1,\ldots,N_n$. Since x^n_i, $x^n_{i+1} \in D \cap B(x_o,r)$, it follows that $u_n(t) \in D \cap B(x_o,r)$ for all t $[t_o,t_o+T]$. Alternatively, the main difficulty in the proof of Theorem 2.1 has been caused by the fact that, without the convexity assumption on $D \cap B(x_o,r)$, the values of u_n may fail to lie in $D \cap B(x_o,r)$. In our present case, $u_n(t) \in D \equiv D(A(t))$ so we can write

$$u_n'(t) - u_m'(t) = A(t)u_n(t) - A(t)u_m(t)$$

$$+ A(t^n_i)x^n_i - A(t)u_n(t) + A(t)u_m(t) - A(t^m_j)x^m_j \qquad (3.18)$$

for $t \in [t^n_i, t^n_{i+1}] \cap [t^m_j, t^m_{j+1}]$. By the properties of u_n (see Theorem 2.2), we have

$$\|A(t^n_i)u^n_i - A(t)u_n(t)\| \leq 1/3n,\ \|A(t)u_m - A(t^m_j)x^m_j\| \leq 1/3m.$$

Then (3.18) yields

$$\frac{d}{dt}\|u_n(t) - u_m(t)\| \leq L(t)\,\|u_n(t)-u_m(t)\| + 1/3m + 1/3m \qquad (3.19)$$

a.e. on $[t_o,t_o + T]$. Solving this elementary inequality (e.g. integrating over $[t_o,t_o + t]$ and then using Gronwall's lemma), we obtain

$$\|u_n(t) - u_m(t)\| \leq T(1/3n + 1/3m)e^{c(t-t_o)},\ t \in [t_o,t_o + T]$$

where $c = \sup\{|L(t)|,\ t \in [t_o,t_o + T]\}$. Thus the existence of $\lim_{n\to\infty} u_n(t)$ uniformly on $[t_o,t_o + T]$ is proved. In other words, the awkward construction of $v_\varepsilon(t)$ (see the proof of Theorem 2.1) is not necessary in the case in which $u_\varepsilon(t) \in D \cap B(x_o,r)$. □

Note that if $D = X$, then Theorem 3.2 yields Theorem 3.7 below. We shall give a second proof for this theorem, independent of the proof of Theorem 3.2. This new proof is interesting in itself.

Theorem 3.7 *Suppose that the function* $(t,x) \to A(t)x$ *is continuous from* $[0,+\infty[\times X$ *into* X *and that there is a continuous function* $L:[0,+\infty[\to [0,+\infty[$ *such that for each* $t \geq 0$, $A(t) - L(t)I$ *is dissipative, i.e.*

$$\langle A(t)x - A(t)y, x-y\rangle_i \leq L(t)\,\|x-y\|^2, \quad t \geq 0,\ x,y \in X.$$

Then, for each $x \in X$, *the problem* $u'(t) = A(t)u, u(0) = x$ *has a unique solution* $u:[0,+\infty[\to X$.

Second proof For simplicity we break the proof of this theorem into two steps. In the first step we assume that $A(t)$ is dissipative (i.e., $L(t) = 0$ for all $t \geq 0$).

Fix $T > 0$ and let $C([0,T];X) \equiv C$ be the space of all continuous functions $u:[0,t] \to X$ with the usual norm $|\cdot|_C$. Define the operator $U:C \to C$ by

$$(Uu)(t) = A(t)u(t), \quad u \in C, \quad t \in [0,T]. \tag{3.20}$$

The dissipativity of $A(t)$ implies

$$\|u(t)-v(t)\| \leq \|u(t)-v(t) - \lambda(A(t)u(t)-A(t)u(t))\|, \quad \forall\lambda > 0$$

for all $u,v \in C$ and $t \in [0,T]$. Therefore $|u-v|_C \leq |u-v-\lambda(Uu-Uv)|_C$, $\forall\lambda > 0$. Thus $U:C \to C$ is continuous and dissipative.

We now define the linear operator $B:D(B) \subset C \to C$ by

$$(Bu)(t) = -u'(t), \quad D(B) = \{u \in C;\ u' \in C,\ u(0) = 0\} \tag{3.21}$$

where u' denotes the strong derivative of u. It is known that B is m-dissipative. According to a theorem of Webb [1], $U + B$ is m-dissipative (i.e., $R(\lambda I-U-B) = C$ for all $\lambda > 0$). In other words, for each $\lambda > 0$ there is $u \in D(B)$ such that

$$\lambda u_\lambda(t) + u_\lambda'(t) = A(t)u_\lambda(t), \quad u_\lambda(0) = 0, \quad \forall t \in [0,T]. \tag{3.22}$$

Let $x^* \in F(u_\lambda(t))$ be such that

$$\langle A(t)u_\lambda(t)-A(t)0, u_\lambda(t)\rangle_i = \langle A(t)u_\lambda(t)-A(t)0,x^*\rangle \leq 0.$$

Then by (3.22) we easily derive

$$\langle u_\lambda'(t),x^*\rangle \leq \langle A(t)0,x^*\rangle \quad \lambda\|u_\lambda(t)\|^2 \leq \|A(t)0\|\ \|u_\lambda(t)\|.$$

Hence

$$\frac{d}{dt}\ \|u_\lambda(t)\| \leq \ A(t)0\| \quad \text{a.e. on } [0,T].$$

This implies $\|u_\lambda(t)\| \leqslant TM$ for all $t \in [0,T]$ and $\lambda > 0$, where $M = \sup \{\|A(t)0\|,\ t \in [0,T]\}$.

Returning to (3.22), we see that

$$u_\lambda'(t)-u_\mu'(t) = A(t)u_\lambda(t)-A(t)u_\mu(t) + \mu u_\mu(t)-\lambda u_\lambda(t),$$

which implies (a.e. on [0,T])

$$\frac{d}{dt}\|u_\lambda(t)-u_\mu(t)\| \leqslant \|\mu u_\mu(t)-\lambda u_\lambda(t)\| \leqslant (\lambda+\mu)MT.$$

Since $u_\lambda(0) = u_\mu(0) = 0$, this inequality yields

$$\|u_\lambda(t)-u_\mu(t)\| \leqslant (\lambda+\mu)MT^2,\ \forall\lambda\ ,\ \mu > 0,\ t \in [0,t].$$

Hence $\lim_{\lambda\downarrow 0} u_\lambda(t) = u(t)$ exists (uniformly on [0,T]). Letting $\lambda \downarrow 0$ in (3.22), we obtain

$$u'(t) = A(t)u(t),\ u(0) = 0 \text{ for all } t \in [0,T].$$

Since $T > 0$ has been arbitrarily chosen, we conclude that the solution u is defined on the whole $[0,+\infty[$.

In the second step one considers the general case (in which $A(t)-L(t)I$ is dissipative). Define $A_i(t):X \to X$ by

$$A_1(t)x = \exp(-\int_0^t L(s)ds)\ A(t)(x \exp \int_0^t L(s)ds),\ x \in X,$$

$$A_2(t)x = A_1(t)x - L(t)x,\ t \geqslant 0,\ \ x \in X.$$

It is easy to check that $A_1(t) - L(t)I$ is dissipative (and consequently $A_2(t)$ is dissipative).

On the basis of the first step there is a unique function $v:[0,+\infty[\to X$ satisfying

$$v'(t) = A_2(t)v(t),\ v(0) = x,\ t \geqslant 0.$$

It follows that the function $u:[0,+\infty[\to X$ defined by $u(t) = v(t) \exp \int_0^t L(s)ds$ is a solution of the problem $u'(t) = A(t)u(t)$, $u(0) = x$. □

In connection with the existence of the solution $u:R \to D$ of the problem

(3.4) we give:

Theorem 3.8 *Assume that* D *is closed and* $A:D \to X$ *is a continuous operator such that*

(1) $|\langle Ax - Ay, x-y\rangle_i| \leq \omega\|x-y\|^2$, $\forall x,y \in D$

for some $\omega > 0$. *Then a necessary and sufficient condition for the problem* $u' = Au$, $u(0) = x \in D$ *to have a solution* $u:R \to D$ *is the following:*

(2) $\lim_{h \to 0} h^{-1}\, d(x + hAx;D) = 0, \quad \forall x \in D.$

Proof Necessity. Since $u(h) \in D$ for $h < 0$, we have

$$\frac{1}{-h}\, d(x + hAx;D) \leq -\frac{1}{h}\, \|x + hAx - u(h)\|$$

$$= \left\| \frac{u(h) - x}{h} - Ax \right\| \to 0 \text{ as } h \uparrow 0.$$

By Theorem 3.3 we already know the necessity of $\lim_{h \downarrow 0} h^{-1}\, d(x + hAx;D) = 0$. Thus the necessity of (2) is proved.

Sufficiency. Set $A_1x = -Ax$, $x \in D$ and consider the problem

$$y'(s) = A_1y(s),\ s \geq 0,\ y(0) = x,\ x \in D. \tag{3.23}$$

In view of Hypothesis (1) we see that A_1 satisfies the conditions of Theorem 3.3. Consequently, (3.23) has a unique solution $y:[0,+\infty[\ \to D$.

It is now clear that the function $v(t) = y(-t)$ $(t \leq 0)$ is the solution to $v' = Av$, $v(0) = 0$ on $]-\infty,0]$. Consequently, the function $u:R \to D$ defined by $u(t) = S(t)x$ if $t \geq 0$, and by $u(t) = y(-t)$ if $t \leq 0$, is the D-valued solution to $u' = Au$, $u(0) = x$, on R. □

Theorem 3.8 yields:

Corollary 3.4 *Let* $D \subset X$ *be closed and let* $A:D \to X$ *be Lipschitz continuous. Then a necessary and sufficient condition for every solution* $u = u(t;t_0,x_0)$ *of the problem* $u'(t) = Au(t)$ *to remain in* D *(i.e.,* $u:R \to D$, $\forall t_0 \in R$, $x_0 \in D$) *is given by the tangential condition* (2) *in Theorem* 3.8.

In terms of flow-invariance (see next section), Corollary 3.4 can be restated as Theorem 4.2, that is:

If $D \subset X$ *is closed and* $A:D \to X$ *is Lipschitz continuous, then the necessary and sufficient condition for the flow-invariance of* D *with respect to* $u' = Au$ *is given by* (3.9).

Theorem 3.7 leads to the non-autonomous version of Theorem 3.8, that is:

Theorem 3.9 *Suppose that the function* $(t,x) \to A(t)x$ *is continuous from* $R \times X$ *into* X *and there exists a continuous function* $L:R \to R_+$ *such that*

$$|\langle A(t)x-A(t)y,x-y\rangle_i| \leq L(t)\,\|x-y\|^2, \quad \forall t \in R, \quad x,y \in X. \tag{3.24}$$

Then for every $(t_0,x_0) \in R \times X$ *the initial value problem* $u' = A(t)u$, $u(t_0) = x_0$, *has a unique solution* $u:R \to X$.

In order to see the generality of Theorem 3.9 with respect to classical existence results, let us recall a theorem due to Bielecki ([1], 1956).

Theorem (Bielecki). *Suppose that there exists a continuous function* $L:R \to R_+$ *such that*

$$\|A(t)x - A(t)y\| \leq L(t)\,\|x-y\|, \quad t \in R, \quad x,y \in X \tag{3.25}$$

$$\|A(t)0\| \leq L(t), \quad t \in R. \tag{3.26}$$

If, in addition, $(t,x) \to A(t)x$ *is continuous from* $R \times X$ *into* X, *then the conclusion of Theorem* 3.9 *holds.*

Clearly, (3.25) implies (3.24). Moreover, in view of Theorem 3.9 it follows that the hypothesis (3.26) is redundant.

§4. Flow-invariance. Applications to component-wise positivity and stability

4.1 Flow-invariant sets with respect to a differential equation. The Nagumo-Brezis theorem

Consider the differential equation

$$u'(t) = A(t)u(t), u(t_0) = x_0, \; t_0 \in \,]a,b[, \; -\infty \leq a < b \leq +\infty. \tag{4.1}$$

Let $\mathbb{A}$ be an open subset of X and let D be a nonempty closed subset of $\mathbb{A}$. For convenience of future reference the following conditions are recorded:

(B1) The function $A:]a,b[\times \mathbb{A} \to X$ is continuous.

(B2) There exists a continuous function $L:]a,b[\to R$ such that

$$\langle A(t)x-A(t)y,x-y\rangle_i \leq L(t) \|x-y\|^2, \quad \forall t \in]a,b[, \ x,y \in A(t).$$

(B3) $\liminf_{h\downarrow 0} h^{-1} d(x+hA(t)x;D) = 0, \quad \forall t \in]a,b[, \ x \in D.$

(B4) $\liminf_{h\to 0} h^{-1} d(x+hA(t)x;D) = 0, \quad \forall t \in]a,b[, \ x \in D.$

(B5) $A:]a,b[\times \mathbb{A} \to X$ is locally Lipschitz (i.e., for each $(t_0,x_0) \in]a,b[\times \mathbb{A}$, there exists a neighbourhood V_0 of (t_0,x_0) and a number $L = L_0 > 0$ with the property $\|A(t)x - A(t)y\| \leq L_0 \|x-y\|$, $\forall (t,x) \in V_0$.

(B6) For each $x_0 \in D \subset \mathbb{A}$, the solution to (4.1) remains in D as long as it exists.

In the (very general) case in which A(t) is defined only on D, Condition (B6) is a problem of existence (see, e.g., Theorem 3.2). If A(t) is defined on $\mathbb{A}$ and D is strictly included in $\mathbb{A}$, then (B6) is a problem of invariance.

Definition 4.1 *The subset* $D \subset \mathbb{A}$ *is said to be a flow-invariant set with respect to the differential equation* (4.1), *if every solution starting in* D *remains in* D *as long as it exists (i.e., if* (B6) *holds).*

On the basis of the results in the previous section, we can easily give a necessary and sufficient condition for the flow-invarinace of D with respect to (4.1).

Theorem 4.1 *Let* D *be a closed subset of* X *and let Hypotheses* (B1) *and* (B2) *be fulfilled. Then* (B3) *is a necessary and sufficient condition for* D *to be a flow-invariant set (to the right) with respect to Equation* (4.1).

Proof The necessity part is a direct consequence of Theorem 3.3. To prove the sufficiency of (B3), let $u:[t_0,T[\to \mathbb{A}$ be the noncontinuable continuation of the local solution to (4.1). Set $\bar{T} = \sup\{c; c > t_0, \ u:[t_0,c[\to D\}$. Clearly, $t_0 < \bar{T} \leq T \leq b$. We will show that $\bar{T} = T$. Indeed, if $\bar{T} < T$, then $\lim_{t\uparrow\bar{T}} u(t) = u(\bar{T})$ since u is continuous on $[t_0,\bar{T}]$. Inasmuch as D is closed, it follows that $u(\bar{T}) \in D$. In view of Theorem 3.1, this contradicts the definition of $\bar{T}$. Therefore the tangential condition (B3) ensures the fact that every $\mathbb{A}$-valued solution (to the right), starting from $x_0 \in D$, remains in D as long as it exists. The proof is complete.

Similarly one proves the following result.

Theorem 4.2 (Nagumo-Brezis). *Suppose that* D *is a closed subset of the Banach space* X *and that* (B1) *and* (B5) *are fulfilled. Then the tangential condition* (B4) *is necessary and sufficient for the flow-invariance of* D *with respect to Equation* (4.1).

In the sequel we give *a second proof of Theorem* 4.2. This second proof is independent of the awkward construction of v_ε in Theorem 2.2. Indeed, by classical Picard's theorem, (B1) and (B5) ensure the existence and uniqueness of the noncontinuable (saturated) solution $u:]T_1, T_2[\to \mathbb{A}$ of (4.1), with $a \leqslant T_1 < t_0 < T_2 \leqslant b$. Let r, M and T be positive numbers such that $B(x_0, r) \subset \mathbb{A}$, $[t_0-T, t_0+T] \subset]T_1, T_2[$, $\|A(t)x\| \leqslant M$ for all $t \in [t_0-T, t_0+T]$ and $\|x-x_0\| \leqslant r$. In addition to these hypotheses, suppose that $T(M+1) \leqslant r$ (which yields $u(t) \in B(x_0, r)$ for $|t-t_0| \leqslant T$) and that A is Lipschitz continuous in the neighbourhood $V_0 = [t_0-T, t_0+T] \times B(x_0, r)$ of (t_0, x_0).

First of all we have

$$u(t) = x_0 + \int_{t_0}^{t} A(s)u(s)ds, \quad t \in]T_1, T_2[.$$

Let us consider the ε-approximate solution u_ε given by (2.24). Consequently,

$$\begin{aligned} u_\varepsilon(t) - u(t) &= \int_{t_0}^{t} (A(s)u_\varepsilon(s) - A(s)u(s))ds \\ &+ \int_{t_0}^{t} (A(a_\varepsilon(s))u_\varepsilon(a_\varepsilon(s)) - A(s)u_\varepsilon(s))ds \\ &+ g_\varepsilon(t), \ t \in [t_0, t_0 + T]. \end{aligned} \tag{4.2}$$

According to the construction of u_ε we have $u_\varepsilon(s) \in B(x_0, r)$ as well as

$$\|A(a_\varepsilon(s))u_\varepsilon(a_\varepsilon(s)) - A(s)u_\varepsilon(s) \leqslant \varepsilon/3, \ s \in [t_0, t_0+T]. \tag{4.2)'$$

Indeed, let $s \in [t_i, t_{i+1}[$. Then $a_\varepsilon(s) = t_i$,

$$|a_\varepsilon(s)-s| = |t_i-s| \leqslant t_{i+1}-t_i, \ \|u_\varepsilon(s)-u_\varepsilon(a_\varepsilon(s))\| \leqslant t_{i+1}-t_i)(M+1).$$

On the basis of Property (5) in Theorem 2.2, we get (4.2). Recall that $\|g_\varepsilon(t)\| \leqslant 2T\varepsilon/3$, for all $|t-t_0| \leqslant T$. Obviously, (4.2) yields

$$\|u_\varepsilon(t) - u(t)\| \leq L \int_{t_0}^{t} \|u_\varepsilon(s) - u(s)\| \, ds + T\varepsilon/3 + 2T\varepsilon/3, \ t \in [t_0, t_0+T].$$

By Gronwall's lemma it follows that

$$\|u_\varepsilon(t) - u(t)\| \leq 2T\varepsilon e^{LT}, \ t \in [t_0, t_0+T],$$

that is, $u(t) = \lim_{\varepsilon \downarrow 0} u_\varepsilon(t)$ uniformly on $[t_0, t_0+T]$. In view of Proposition 2.1, $\lim_{\varepsilon \downarrow 0} u_\varepsilon(t) \in D$ for $t \in [t_0, t_0+T]$, hence $u(t) \in D$ also on $[t_0, t_0+T]$. Arguing as in the proof of Theorem 4.1, one shows that $u(t) \in D$ for all $t \in [t_0, T_2[$. Similarly we prove that $u(t) \in D$ for $t \in]T_1, t_0]$.

Let us discuss the case $X = R^m$ (the Euclidean space). The following result holds:

Theorem 4.3 *Let* D *be a closed subset of* R^m. *Suppose that* (B1) *holds (with* $X = R^m$). *Then* (B4) *is a necessary and sufficient condition in order that for each* $t_0 \in]a,b[$ *and* $x_0 \in D$ *at least one solution (to the right) of* (4.1) *satisfies* $u(t) \in D$ *on some intervals* $[t_0, t_0 + \bar{T}[$, *with* $0 < \bar{T} \leq b - t_0$.

Proof In view of Proposition 2.1 it suffices to show that the approximate solution u_ε (given by (2.24)) contains a convergent subsequence. This is true in this case, by the Ascoli-Arzela lemma. Indeed, we have already seen that $\|u_\varepsilon(t)-u_\varepsilon(s)\| \leq |t-s| (M + 2\varepsilon/3)$ and $u_\varepsilon(t) \in B(x_0, r)$ for all $t,s \in [t_0, t_0 + T]$. On the basis of Proposition 2.1, $\lim_{\varepsilon \downarrow 0} u(t) = u(t)$ is a D-valued solution to (4.1) on $[t_0, t_0 + T]$. By Zorn's lemma, there is a maximal interval $[t_0, t_0 + \bar{T}[\subset [t_0, b[$ such that $u(t) \in D$ for all $t \in [t_0, t_0 + \bar{T}[$ (see the proof of Proposition 2.2).

Lemma 4.1 *Let us write* $A(t)x$ *in the form*

$$A(t)x = (f_1(t, x_1, \ldots, x_m), \ldots, f_m(t, x_1, \ldots, x_m)), \ t \in]a,b[,$$

$x = (x_1, \ldots, x_m) \in \mathbb{A} \subset R^m$. *Let* $a_i < b_i$, $i = 1, \ldots, m$. *Set*

$$D_0 = \prod_{i=1}^{m} [a_i, b_i] = \{x \in R^m, \ a_i \leq x_i \leq b_i, \ i = 1, \ldots, m\}.$$

Assume that $D_0 \subset \mathbb{A}$. *Then* (B3) *is equivalent to*

$$(B3)' \begin{cases} f_i(t,x_1,\ldots,x_{i-1}, a_i, x_{i+1},\ldots,x_m) \geqslant 0 \\ f_i(t,x_1,\ldots,x_{i-1}, b_i, x_{i+1},\ldots,x_m) \leqslant 0 \end{cases}$$

for all $t \in]a,b[$, $x \in D_0$, $i = 1,2,\ldots,m$ *with the convention* $(x_0 = x_1, x_{m+1}=x_m)$.

<u>Proof</u> We know that (B3) is equivalent to

$$x + h(A(t)x + r(h)) \in D, \quad \forall t \in]a,b[, \quad x \in D \tag{4.3}$$

for some $r:R_+ \to R^m$ with $r(h) \to 0$ as $h \downarrow 0$ (see (3.9')).

In this case (4.3) becomes

$$a_i \leqslant x_i + h(f_i(t,x_1,\ldots,x_m) + r_i(h)) \leqslant b_i, \; h > 0 \tag{4.4}$$

for all $a_i \leqslant x_i \leqslant b_i$, $t \in]a,b[$ and $i = 1,2,\ldots,m$, where $r(h) = (r_1(h),\ldots$ $\ldots,r_m(h)) \to 0$ as $h \downarrow 0$. Obviously (B3)' implies (4.4) with $r_i(h) = 0$. Conversely, since $a_i < b_i$, and $r_i(h) \to 0$ as $h \downarrow 0$, we see that (4.4) implies (B3)'. □

Combining Theorem 4.2 and Lemma 4.1 one obtains:

<u>Corollary 4.1</u> *Suppose that* $f_i:]a,b[\times \mathbb{A} \to R^m$ *are continuous and locally Lipschitz with respect to* x (*in the sense of hypothesis* (B5)) *for each* $i = 1,\ldots,m$. *Then* D_0 *is a flow-invariant set (to the right) with respect to the system*

$$x_i'(t) = f_i(t,x_1,\ldots,x_m), \; x_i(t_0) = x_i^0 \tag{4.5}$$

if and only if (B3)' (*in Lemma* 4.1) *is fulfilled.*

4.2 <u>Component-wise positive solutions</u>. <u>Invariant rectangles</u>

It is now easy to get necessary and sufficient conditions for the existence of positive solutions to (4.5). Set

$$R_+^m = \{x = (x_1,\ldots,x_m) \in R^m; \; x_i \geqslant 0, i = 1,\ldots,m\} \tag{4.6}$$

and assume that $R_+^m \subset \mathbb{A}$. Arguing as in the proof of Lemma 4.1 (with $a_i = 0$), one easily obtains:

Corollary 4.2 *Suppose that the hypotheses of Corollary 4.1 are fulfilled. Then for every* $t_0 \in]a,b[$ *and* $x_0 = (x_1^0,\ldots,x_m^0) \in R_+^m$, *each component* x_i *of the solutions to* (4.5) *is positive (i.e.,* R_+^m *is a flow-invariant set to the right for* (4.5)) *if and only if*

$$f_i(t,x_1,\ldots,x_{i-1},0,\ x_{i+1},\ldots,x_m) \geq 0, \tag{4.7}$$

$i = 1,\ldots,m$ *(with the convention* $x_0 = x_1, x_{m+1} = x_m$), *for all* $t \in]a,b[$ *and* $x \in R_+^m$.

Let us consider the linear case,

$$f_i(t,x_1,\ldots,x_m) = a_{i1}(t)x_1 + \ldots + a_{im}(t)x_m, \tag{4.8}$$

$i = 1,\ldots,m$. In this case (4.7) is obviously equivalent to

$$a_{ij}(t) \geq 0,\ j = 1,2,\ldots\ i-1,\ i+1,\ldots,m,\ t \in]a,b[. \tag{4.9}$$

Consequently, if we denote by $\bar{A}(t) = (a_{ij}(t))$, $i,j = 1,\ldots,m$ (the $m \times m$ matrix with the elements $a_{ij}(t)$), we can state the classical result.

Corollary 4.3 *Let us consider the* $m \times m$ *matrix* $A(t) = (a_{ij}(t))$. *Suppose that each element* a_{ij} *is a continuous function from* $]a,b[$ *into* R. *Then a necessary and sufficient condition in order that, for every initial condition* $x_0 = (x_1^0,\ldots,x_m^0)$ *with* $x_0 \in R_+^m$, *each component* x_i *of the solution* x *to the problem* $x'(t) = A(t)x(t)$, $x(t_0) = x_0$, *satisfies* $x_i(t) \geq 0$, *for all* $t \in [t_0,b[$, *is*

$$a_{ij}(t) \geq 0,\ \textit{for all}\ t \in [t_0,b[,\quad i \neq j;\quad i,j = 1,2,\ldots,m.$$

We now proceed to extend the result given by Lemma 4.1 to the time-dependent case. Precisely, let us consider the case

$$D(t) = \prod_{i=1}^{m} [a_i(t),b_i(t)] = \{x \in R^m,\ a_i(t) \leq x_i \leq b_i(t),\ i = 1,\ldots,m\}$$

$$a_i(t) < b_i(t),\quad t \in]a,b[. \tag{4.10}$$

Lemma 4.2 *Assume that* a_i *and* b_i *are differentiable functions from* $]a,b[$ *into* R. *Then (with the same notations as in Lemma* 4.1) *Condition* (A4) *in* §2 *is equivalent to*

$$\left.\begin{array}{l} f_i(t,x_1,\ldots,x_{i-1},a_i(t), x_{i+1}),\ldots,x_m) \geq a_1'(t) \\ f_i(t,x_1,\ldots,x_{i-1},b_i(t), x_{i+1},\ldots,x_m) \leq b_i'(t) \end{array}\right\} \tag{4.11}$$

for all $a_j(t) \leq x_j \leq b_j(t)$, $j = 1,\ldots,i-1, i+1,\ldots,m, t \in]a,b[$ *and* $i = 1,\ldots,$ *where* a_1' (b_i') *denotes the derivative of* a_i (*resp.* b_i).

Proof In this case (A4) is equivalent to

$$x + h(A(t)x + r(h)) \in D(t+h),\ x \in D(t),\ t \in]a,b[,\ h > 0 \tag{4.12}$$

for some $r:R_+ \to R^m$ with $r(h) \to 0$ as $h \downarrow 0$. Clearly, $r = r(h,x)$ depends also on x and (4.12) means

$$a_i(t+h) \leq x_i + hf_i(t,x_1,\ldots,x_m) + hr_i(h) \leq b_i(t+h),\ h > 0 \tag{4.13}$$

for $a_i(t) \leq x_i \leq b_i(t)$, $t \in]a,b[$, $i = 1,\ldots,m$.

The differentiability of a_i and b_i is equivalent to the existence of the functions r_i^1, r_i^2 having the properties

$$\left.\begin{array}{l} a_i(t + h) - a_i(t) = ha_i'(t) + hr_i^1(h) \\ b_i(t + h) - b_i(t) = hb_1'(t) + hr_i^2(h) \end{array}\right\} \tag{4.13$'$}$$

and $r_i^p(h) \to 0$ as $h \downarrow 0$, $p = 1,2$, $t \in]a,b[$.

It is now easy to check that (in view of (4.13)' the conditions (4.11) are equivalent to (4.13).

For D(t) given by (4.10), Hypotheses (A1)-(A3) are fulfilled. Then, similarly to Corollary 4.1 (using Theorem 2.1 Lemma 4.2, the convergence of u_ε in the case of R^m and Proposition 2.1), we derive:

Corollary 4.4 *Suppose that the functions* $(t,x) \to f(t,x)$, $x \in D(t)$ *are continuous and the hypotheses of Lemma* 4.2 *are fulfilled. Then* (4.11) *are necessary and sufficient conditions in order that for every* $t_0 \in]a,b[$ *and* $x_0 = (x_1^0,\ldots,x_m^0) \in D(t_0)$ (*i.e.*, $a_i(t_0) \leq x_i^0 \leq b_i(t_0)$) *at least one solution* x *to* (4.5) *satisfies* $a_i(t) \leq x_i(t) \leq b_i(t)$ *on its maximal interval of existence (to the right of* t_0*).*

4.3 Validation of a system modelling enzymatic reactions

Let us consider the classical enzymatic reaction

$$E + S \rightleftarrows C \to E + P$$

where E, S, C and P denote enzyme, substrate, complex and product, respectively. The evolution of the substrate and complex concentrations (denoted by x_1 and x_2 respectively) for large values of the temporal variable, t, is described by the following parameter differential system (see, e.g., J.D. Murray [1, p.9] and S.I. Rubinov [1, p.52]:

$$\left.\begin{aligned} x_1' &= -x_1 + (x_1 + a)x_2,\ x_1(t_o) = x_1^o \\ \mu x_2' &= x_1 - (x_1 + b)x_2,\ x_2(t_o) = x_2^o, t_o \geqslant 0 \end{aligned}\right\} \tag{4.14}$$

where $\mu > 0$ is a small parameter and $0 < a \leqslant b$. We shall give a mathematical explanation of the fact that (under natural assumptions on the evolution of x_1 and x_2) the condition $0 < a \leqslant b$ is necessary. Precisely, it is natural to assume that there are $b_i > 0$, $i = 1,2$, such that the solution $x = (x_1(t), x_2(t))$ of (4.14) is positive and bounded on its right maximal interval of existence, as follows:

$$0 \leqslant x_1(t) \leqslant b_1,\ 0 \leqslant x_2(t) \leqslant b_2,\ t \in [t_o, T[, \tag{4.15}$$

whenever the initial state $x_o = (x_1^o, x_2^o)$ satisfies $0 \leqslant x_1^o \leqslant b_1$, $0 \leqslant x_2^o \leqslant b_2$. This natural assumption on the evolution of $x_1(t)$ and $x(t)$ is mathematically equivalent to the flow-invariance (to the right) of the closed set. $D_o = [0,b_1] \times [0,b_2]$ with respect to the (nonlinear) system (4.14). In this case,

$$f_1(x_1,x_2) = -x_1 + (x_1 + a)x_2$$

$$f_2(x_1,x_2) = \mu^{-1}(x_1 - x_1x_2 - bx_2),$$

and so the hypotheses of Corollary 4.1 are obviously satisfied. Therefore the necessary and sufficient conditions for the conditions (4.15) to be satisfied are given by (B3)' (see Lemma 4.1). In this case $a_1 = a_2 = 0$, so (B3)' becomes

$$f_1(0,x_2) = ax_2 \geqslant 0,\quad f_2(x_1,0) = \mu^{-1}x_1 \geqslant 0 \tag{4.16}$$

$$f_1(b_1,x_2) = - b_1 + (b_1 + a)x_2 \leqslant 0,$$

$$f_2(x_1,b_2) = \mu^{-1}(x_1 - x_1b_2 - bb_2) \leqslant 0 \qquad (4.17)$$

for all $0 \leqslant x_1 \leqslant b_1$ and $0 \leqslant x_2 \leqslant b_2$.

It is clear that (4.16) is equivalent to $a \geqslant 0$. For practical reasons, take $a > 0$. The first condition in (4.17) is equivalent to $(b_1+a)b_2 \leqslant b_1$, i.e., to $a \leqslant ((1-b_2)b_1)/b_2$. Therefore, necessarily, $b_2 \leqslant 1$. Moreover, $0 < a$ implies $b_2 < 1$. The second condition in (4.17), i.e., $(1-b_2)x_1 \leqslant bb_2$, $\forall 0 \leqslant x_1 \leqslant b_1$, is equivalent to $(1-b_2)b_1 \leqslant bb_2$, i.e., to $b \geqslant ((1-b_2)b_1)/b_2$. Consequently, the following result has been proved:

Proposition 4.1 *Let* $b_i > 0$ $(i = 1,2)$ *and* $\mu > 0$. *The conditions*

$$0 \leqslant a \leqslant \frac{(1 - b_2)b_1}{b_2} \leqslant b, \quad b_2 \leqslant 1 \qquad (4.18)$$

are necessary and sufficient in order for the evolution of x_1 *and* x_2 *to satisfy* (4.15).

Remark 4.1 From (4.18) we see that if $0 < a$, then necessarily $b_2 < 1$ (so $0 < b_2 < 1$). In other words, if $0 < a$, then the flow-invariance of $D_o = [0,b_1] \times [0,b_2]$ for system (4.14) is impossible if $b_2 \geqslant 1$. Moreover, D_o is flow-invariant to the right for (4.14) iff (4.18) holds. The fact that $0 \leqslant x_2(t) \leqslant b_2 < 1$ is very natural, since the complex concentration $x_2(t)$ cannot be greater than 1. Therefore we now have also a mathematical explanation of the fact that the second component x_2 of the solution (x_1,x_2) to (4.14) is [0,1]-valued. In other words, we have proved that all flow-invariant sets (of the form $D_o = [0,b_1] \times [0,b_2]$) with respect to (4.14) are given by D_o with b_2 arbitrarily choosen in]0,1[and

$$ab_2(1-b_2)^{-1} \leqslant b_1 \leqslant bb_2(1-b_2)^{-1}. \qquad (4.19)$$

Now let us characterize the rectangles

$$D_1 = [a_1,b_1] \times [a_2,b_2], \; 0 < a_i < b_i, \; i = 1,2 \qquad (4.20)$$

which are invariant with respect to (4.14). In this case, the necessary and sufficient conditions (B3)' in Lemma 4.1 become (4.17) and (4.21) below:

$$0 \leqslant -a_1 + (a_1 + a)x_2,\ a_2 \leqslant x_2 \leqslant b_2,$$
$$0 \leqslant x_1 - (x_1 + b)a_2,\ \ a_1 \leqslant x_1 \leqslant b_1. \tag{4.21}$$

Clearly, the conditions (4.21) are equivalent to

$$a_1(a_1 + a)^{-1} \leqslant a_2 \leqslant a_1(a_1 + b)^{-1} \tag{4.22}$$

which implies also $a \geqslant b$. In view of (4.18) it follows that $a = b$. Finally, for $a = b$, (4.18) and (4.22) give, respectively,

$$a_2 = a_1(a_1 + a)^{-1},\ b_2 = b_1(b_1 + a)^{-1} \tag{4.23}$$

with arbitrary $a_1 > 0$, $b_1 > 0$.

To sum up, we have got the following result:

Proposition 4.2 *The necessary and sufficient condition for system* (4.14) *to admit an invariant set of the form* (4.20) *is* $a = b$ (*where* $a > 0$). *All rectangles of the form* (4.20) *which are invariant sets with respect to* (4.14) *are given by* (4.23). *In particular, the rectangle* $[a,2a] \times [1/2, 2/3]$ *is a flow invariant set with respect to* (4.14).

4.4 Component-wise positive asymptotic stability (CWPAS). Special Hurwitz metrices

In this subsection we are concerned with the flow invariance of t-dependent sets given by (4.10) with $a_i(t) = 0$, i.e.,

$$D(t) = \prod_{i=1}^{m} [0,b_i(t)],\ 0 < b_i(t),\ \ \forall t \geqslant 0;\ i = 1,\dots,m \tag{4.24}$$

where b_i are continuously differentiable on $]a,b[$ (or on $R_+ = [0,+\infty[$).

As in Definition 4.1, $D(t)$ is said to be a flow-invariant (or simply, invariant) set with respect to system (4.5) if, for each $t_0 \geqslant 0$ and x_0 of components x_i^0 with $x_0 \in D(t_0)$, the corresponding solution x to the system (4.5) remains in $D(t)$ (i.e., $x(t) \in D(t)$) as long as it exists. In other words, the flow invariance of $D(t)$ with respect to (4.5) means that, for each initial condition $x(t_0) = x_0$ with $0 \leqslant x_0 \leqslant b(t_0)$, the corresponding solution x to (4.5) satisfies

$$0 \leqslant x(t) \leqslant b(t), \quad \forall t \geqslant t_o. \tag{4.25}$$

Denote by $b(t)$ the m by 1 vector of components $b_i(t)$, $i = 1,\dots,m$. Let v be another m by 1 vector of components v_i, $i = 1,\dots,m$. We say that $b(t) > v(t)$, $b(t) \geqslant v(t)$ iff $b_i(t) > v_i(t)$ and $b_i(t) \geqslant v_i(t)$, $i = 1,\dots,m$, respectively. The vector of components zero is also denoted by 0.

On the basis of Lemma 4.2 and Corollary 4.4 we have

<u>Proposition 4.3</u> *Suppose that* $f_i :]a,b[\times A \to R^m$ *are continuous and locally Lipschitz with respect to* x. *Then the necessary and sufficient conditions for the flow invariance of* D(t) *defined by* (4.24)) *are given by* (4.11) *with* $a_i(t) = 0$, *for all* $t \in]a,b[$ *and* $i = 1,\dots,m$.

Now let us treat the case in which (4.5) is the linear homogeneous system with constant coefficients

$$x'(t) = Ax(t) \tag{4.26}$$

where $A = (a_{ij})$, $a_{ij} \in R$, $i,j = 1,\dots,m$.

In view of Corollary 4.3, a necessary condition for the invariance of D(t) with respect to (4.26) is given by

$$a_{ij} \geqslant 0, \; i \neq j, \; i,j = 1,\dots,m. \tag{4.27}$$

Furthermore, taking into account Corollary (4.4) we easily get:

<u>Proposition 4.4</u> *The necessary and sufficient conditions for the flow-invariance of* D(t) *defined by* (4.24) *are the following:*

(1) $a_{ij} \geqslant 0$, $i = j$, $i,j = 1,\dots,m$.

(2) $b(t)$ *satisfies the inequation*

$$b'(t) \geqslant Ab(t), \quad \forall t \geqslant 0. \tag{4.28}$$

In the particular case $b(t) = v = \text{const.}$, where v has the components v_i, it follows that the set $D = \prod_{i=1}^{m} [0,v_i]$ with $v_i > 0$, is invariant with respect to (4.26), *iff* the conditions (4.27) and $Av \leqslant 0$ hold.

In what follows, in addition to $b_i(t) > 0$, we introduce the stability condition

$$\lim_{t\to\infty} b_i(t) = 0, \; i = 1,\ldots,m. \tag{4.29}$$

Definition 4.2 *The system* $x' = Ax$ *is said to be component-wise positive asymptotically stable (in short:* (CWPAS)), *if there is a vector* $b(t) > 0$ *satisfying* (4.29) *such that the set* $D(t)$ *defined by* (4.24) *is flow-invariant with respect to this system (i.e.,* (4.25) *holds).*

For such a vector $b(t)$ we say that $x' = Ax$ is (CWPAS) with respect to $b(t)$. One assumes that $t \to b(t)$ is continuously differentiable on $R = [0,+\infty[$.

Remark 4.2 Let e^{At} be the fundamental matrix of system (4.26). Suppose that A satisfies (4.27). Then all elements of e^{At} are non-negative. It follows that for each $b(t) > 0$, each component of $e^{At}b(t)$ is strictly positive (i.e., $e^{At}b(t) > 0$).

What properties of A guarantee the component-wise asymptotic stability? The complete answer to this question is given by:

Theorem 4.4 *The system* $x' = Ax$ *is* (CWPAS) *iff* A *satisfies the following three conditions:*

(1) $a_{ij} \geq 0$, $i \neq j$, $i,j = 1,\ldots,m$.

(2) $a_{ii} < 0$, $i = 1,\ldots,m$.

(3) A *has the Hurwitz property (i.e., the real part of each eigenvalue of* A *is negative).*

Proof Necessity. We have already seen that Property (1) is equivalent to non-negativity of the solutions (Corollary 4.3). In viee of (4.29) there are $t_i > 0$ such that $b'(t_i) < 0$. Consequently, for $t = t_i$ (4.28) yields

$$0 > a_{ii}b_i(t_i) + \prod_{j=1}^{m} a_{ij}b_j(t), \; j \neq i; \; i = 1,\ldots,m \tag{4.30}$$

which implies (2). Clearly, (CWPAS) implies asymptotic stability in the classical sense of (4.26), and therefore (3) is also necessary.

Sufficiency. If Conditions (1)-(3) are fulfilled, then according to Remark 4.2 and (4.28), system (4.26) is (CWPAS) with respect to $b(t) = e^{At}x_o$, for each m by 1 vector $x_o > 0$. The proof is complete.

We now prove that (CWPAS) is equivalent to the existence of some m by 1 vectors v of components $v_i > 0$, and some numbers $c > 0$ such that (4.26) is (CWPAS) with respect to $b(t) = ve^{-ct}$. In this case we say that system (4.26) is $(CWPAS)_{cv}$. On the basis of (4.30) and of Proposition 4.4 we have:

Theorem 4.5 *The system* $x' = Ax$ *is* $(CWPAS)_{cv}$ *iff the following three conditions hold:*

(1) $a_{ij} \geqslant 0, \forall i \neq j, i,j = 1,\ldots,m.$

(2) $a_{ii} < 0, \quad i = 1,\ldots,m.$

(3) $Av < 0, \; c \leqslant \min\limits_{i=\overline{1,m}} \{-a_{ii} - \frac{1}{v_i} \sum\limits_{j=1}^{m} a_{ij}v_j; \; j \neq i\}.$

Remark 4.3 From Theorems 4.4 and 4.5 it follows that, under Hypotheses (1) and (2) of Theorem 4.5, the matrix A has the Hurwitz property if and only if there are some m by 1 vectors $v > 0$ such that $Av < 0$. A very simple and useful condition which is equivalent to the Hurwitz property is given by:

Theorem 4.6 *Let* $A = (a_{ij})$ *be a* m *by* m *matrix with the properties:*

(P) $\quad a_{ii} < 0, \; a_{ij} \geqslant 0, \; i \neq j; \; i,j = 1,\ldots,m.$

Then a necessary and sufficient condition for the existence of a vector $v > 0$ *such that* $Av < 0$ *(or equivalently: for* A *to have the Hurwitz property) is that*

$$d^{(k)}_{p,p} > 0, \; k = 1,\ldots,m-1; \; p = 1,\ldots,m-k \tag{4.31}$$

where $d^{(k)}_{p,q}$ *are inductively defined by*

$$d^{(0)}_{i,j} = a_{ij}, \; d^{(k)}_{p,q} = \begin{vmatrix} d^{(k-1)}_{p,q} & d^{(k-1)}_{p,m-k+1} \\ d^{(k-1)}_{m-k+1,q} & d^{(k-1)}_{m-k+1,m-k+1} \end{vmatrix} \tag{4.32}$$

$i,j = 1,\ldots,m; \; k = 1,\ldots,m-1; \; p,q = 1,\ldots,m-k.$

Proof Under Hypothesis (P) we have to characterize the solutions $v_i > 0$ to the system of inequalities

$$a_{i1}v_1 + a_{i2}v_2 + \dots + a_{im}v_m < 0,\ i = 1,\dots,m. \tag{4.33}$$

Since $a_{mm} < 0$ and $a_{pm} \geqslant 0$, $\forall p = 1,\dots,m-1$, (4.33) is equivalent to

$$\frac{a_{m1}}{-a_{mm}} v_1 + \dots + \frac{a_{mm-1}}{-a_{mm}} v_{m-1} < v_m < \frac{-a_{p1}}{a_{pm}} v_1 + \dots + \frac{-a_{pm-1}}{a_{pm}} v_{m-1}, \tag{4.34}$$

$p = 1,\dots,m-1$.

Possible strictly positive values for v_m will consequently exist if and only if $v_p > 0$ satisfy the m-1 by m-1 system

$$-d^{(1)}_{p,1}v_1 - \dots - d^{(1)}_{p,m-1}v_{m-1} < 0,\ p = 1,\dots,m-1. \tag{4.35}$$

Since

$$d^{(1)}_{p,q} = \begin{vmatrix} a_{pq} & a_{pm} \\ a_{mq} & a_{mm} \end{vmatrix} \quad p,q = 1,\dots,m-1, \tag{4.36}$$

we have $d^{(1)}_{p,q} \leqslant 0$, for $p \neq q$. Accordingly, for the existence of $v_p > 0$, $p = 1,\dots,m-1$ satisfying (4.35), the conditions $d^{(1)}_{p,p} > 0$ are necessary, $\forall p = 1,\dots,m$.

Now let us observe that the matrix of system (4.35) also satisfies Condition (P) with $d^{(1)}_{p,q}$ in place of a_{pq}. Consequently, we proceed to eliminate v_{m-1} from system (4.35) as v_m was eliminated from (4.33), and so on. By successive application of this process of elimination one obtains

$$\frac{-d^{(k)}_{m-k,1}}{d^{(k)}_{m-k,m-k}} v_1 + \dots + \frac{-d^{(k)}_{m-k,m-k-1}}{d^{(k)}_{m-k,m-k}} v_{m-k-1} < v_{m-k} <$$

$$\frac{d^{(k)}_{p,1}}{-d^{(k)}_{p,m-k}} v_1 + \dots + \frac{d^{(k)}_{p,m-k-1}}{-d^{(k)}_{p,m-k}} v_{m-k-1},\quad p = 1,\dots,m-k-1,\ k = 1,\dots,m-2, \tag{4.37}$$

and the necessity of Conditions (4.31).

After the elimination of $v_m, v_{m-1},\dots,v_3$, the remaining variables v_1, v_2 necessarily satisfy (4.37) with $k = m-2$, i.e.,

$$\frac{-d_{2,1}^{(m-2)}}{d_{2,2}^{(m-2)}} v_1 < v_2 < \frac{d_{1,1}^{(m-2)}}{-d_{1,2}^{(m-2)}} v_1 \tag{4.38}$$

with $d_{1,2}^{(m-2)}, d_{2,1}^{(m-2)} \leq 0$; $d_{1,1}^{(m-2)}, d_{2,2}^{(m-2)} > 0$.

In these conditions, inequalities (4.38) with $v_1, v_2 > 0$ are equivalent to

$$\begin{vmatrix} d_{1,1}^{(m-2)} & d_{1,2}^{(m-2)} \\ d_{2,1}^{(m-2)} & d_{2,2}^{(m-2)} \end{vmatrix} v_1 \equiv -d_{1,1}^{(m-1)} v_1 < 0 \tag{4.39}$$

and therefore the condition $d_{1,1}^{(m-1)} > 0$ is necessary and sufficient for the existence of some $v_1 > 0$ satisfying (4.39). Clearly, in (4.39) the variable v_1 may be assigned positive values at pleasure. Next we take a v_2 satisfying (4.38). In such a manner, going back from (4.39) to (4.37) we prove the existence of $v_1, \ldots, v_{m-1}$ (with $v_1 > 0$, arbitrary) satisfying (4.35), and the necessity of (4.31). The remaining variable v_m is merely required to satisfy inequalities (4.34). This completes the proof.

4.5 <u>(CWPAS) of a system modelling the spatial spread of a class of bacterial and viral diseases</u>

The following system was proposed to model the cholera epidemic which spread in European Mediterranean regions in 1973:

$$\begin{aligned} x_1' &= -a_{11}x_1 + a_{12}x_2 \\ x_2' &= g(x_1) - a_{22}x_2 \end{aligned} \qquad t \geq 0 \tag{4.40}$$

where x_1 is the average concentration of bacteria, x_2 is the infective human population, $-a_{11}x_1$ describes the natural growth rate of bacteria, $a_{12}x_2$ the contribution of the infective humans to the growth rate of bacteria and $-a_{22}x_2$ the natural damping of the infective human population due to finite mean duration of the infectiousness of humans. Finally, $g(x_1)$ describes the infection rate of humans under the assumption that the total susceptible human population is constant.

It is hoped that similar systems can be used to model other epidemics (e.g. typhoid fever, poliomyelitis, hepatitis).

A more realistic model seems to be the reaction-diffusion system

$$\begin{aligned} \frac{\partial u_1}{\partial t} &= d_1 \ \Delta u_1 - a_{11}u_1 + a_{12}u_2 \\ & \qquad\qquad\qquad\qquad , \quad t \geq 0,\ x \in \Omega \\ \frac{\partial u_2}{\partial t} &= d_2 \ \Delta u_2 + g(u_1) - a_{22}u_2 \end{aligned} \tag{4.41}$$

with the mixed conditions

$$\frac{\partial u_i}{\partial n} \in -\partial\beta_i(u_i) \text{ on }]0,\infty[\times \Gamma , \quad u_i(0,x) = u_i^0(x) \text{ on } \Omega, \quad i = 1,2. \tag{4.42}$$

Here u_1 and u_2 denote respectively the spatial concentration of bacteria and of human population in the bounded domain (habitat) $\Omega \subset R^N$, with smooth boundary Γ. As usual, Δ denotes the Laplace operator in R^N. The diffusion coefficients d_1 and d_2 are non-negative, $\partial/\partial n$ is the outward normal derivative, while $\partial\beta_i$ is the subdifferential of the lower semicontinuous convex function $\beta_i : R \to [0,+\infty]$, $i = 1,2$. A basic hypothesis is that the bacteria diffuse randomly in Ω, while the random diffusion of the human population may be neglected with respect to that of bacteria (i.e., we can also take $d_2 = 0$).

First we are concerned with the componentwise positive asymptotic stability of system (4.40), in the sense of Definition 4.2. As in the case of Theorem 4.5, we can easily find necessary and sufficient conditions for the existence of some positive numbers v_1, v_2 and c such that system (4.40) is $(\text{CWPAS})_{cv}$. Namely, the following result holds:

<u>Proposition 4.5</u> *Suppose that* $g : R \to R$ *is continuous and guarantees the uniqueness of the solution to the Cauchy problem for* (4.40). *Then for each non-negative initial condition* x_0 *of components* x_1^0, x_2^0 *the corresponding solution to* (4.40) *is non-negative (i.e.,* $R_+ \times R_+$ *if flow-invariant) iff*

(1) $a_{12} \geq 0$, $g(x_1) \geq 0$, $\forall x_1 \geq 0$.

Assume in addition that

(1^0) $d \equiv \sup\limits_{0<s\leq v_1} g(s)/s < +\infty$, $g : R_+ \to R_+$.

Then the rectangle $D(t) = [0,v_1e^{-ct}] \times [0,v_2e^{-ct}]$, $t \geq 0$, *is flow-invariant with respect to* (4.40) (*i.e.*, (4.40) *is* $(\text{CWPAS})_{cv}$ *in the sense of* (4.25)) *iff*

$$(2)\quad \begin{vmatrix} a_{11} & a_{12} \\ d & a_{22} \end{vmatrix} > 0,\ a_{11} > 0,\ a_{22} > 0,$$

$$(3)\quad c \leq \min\{a_{11} - a_{12}v_2/v_1,\ a_{22} - dv_1/v_2\}.$$

Proof In this case, Conditions (4.7) become (1). For the second part one applies Corollary 4.4. Clearly, in the case of (4.40), conditions (4.11) with $a_1 = a_2 = 0$ and $b_1(t) = v_1e^{-ct}$, $b_2 = v_2e^{-ct}$ are equivalent to

$$-a_{11}v_1 + a_{12}v_2 \leq -dv_1,\ g(v_1e^{-ct}) - a_{22}v_2e^{-ct} \leq -cv_2e^{-ct} \qquad (4.43)$$

for all $t \geq 0$, where $v_1, v_2, c > 0$. Since the function $t \to v_1e^{-ct}$ from $[0,+\infty[$ into $]0,v_1]$ is surjective, it follows that Conditions (4.43) are equivalent to (2) and (3). This completes the proof.

Remark 4.3 In the hypotheses of Proposition 4.5, we can also assert that (4.40) is $(\text{CWPAS})_{cv}$ *iff* $v_1 > 0$ is arbitrary,

$$\frac{d}{a_{22}} < \frac{v_2}{v_1} < \frac{a_{11}}{a_{12}},\ a_{11}, a_{22} > 0 \text{ and } c \text{ satisfies (3).} \qquad (4.44)$$

In particular, if $g(x_1) = a_{21}x_1$, then $d = a_{21}$, and this proposition is just Theorem 4.5 with $m = 2$. In other words, the evolution of the average concentration of bacteria (described by $x_1(t)$) and the infective human population $x_2(t)$ has the behaviour (exponential decay)

$$0 \leq x_1(t) \leq v_1e^{-ct},\ 0 \leq x_2(t) \leq v_2e^{-ct},\quad \forall t \geq 0 \qquad (4.45)$$

with $c, v_1, v_2 > 0$, if and only if Conditions (1), (2) and (3) of Proposition 4.5 hold, or equivalently, *iff* (1) and (4.44) hold.

The existence of a global solution of the problem (4.41) + (4.42), is given by Theorem 5.9 in Chapter 5.

4.6 Positivity and flow invariance with respect to some systems modelling chemical reactions

The quantitative model for the chemical mechanism in the Belousov reaction has been given by Field and Noyes [1]. Their model consists of five irreversible steps, namely

$$A + Y \to X;\ X + Y \to P;\ B + X \to 2X + Z;\ 2X \to Q;\ Z \to fY. \tag{4.46}$$

Here f is a stoichiometric factor, P and Q are products and

$$X = HBrO_2,\ Y = Br^-,\ Z = Ce(IV),\ A = B = BrO_3^-.$$

If we denote by x, y, z, a and b the concentrations of the quantities X, Y, Z, A and B respectively, then the kinetic system for (4.46) is

$$\begin{aligned} x' &= k_1ay - k_2xy + k_3bx - 2k_4x^2, \\ y' &= -k_1ay - k_2xy + fk_5z, \\ z' &= k_3bx - k_5z, \end{aligned} \tag{4.47}$$

where the constants f and k_i (i = 1,2,3,4,5) are positive. By an appropriate change of variables this system can be written in the form

$$\begin{aligned} x_1' &= s(x_2 - x_1x_2 + x_1 - qx_1^2), \\ x_2' &= s^{-1}(fx_3 - x_2 - x_1x_2), \\ x_3' &= w(x_1 - x_3), \end{aligned} \tag{4.48}$$

where s, w and q are positive constant. In what follows it is important to point out that q is less than one. Precisely, $q = 8.375 \times 10^{-6}$.

In view of the significance of (4.47), or equivalently (4.48), we have to prove that the positive octant $x_1, x_2, x_3 \geqslant 0$, (denoted also by R_+^3, see (4.6)) is flow-invariant with respect to each of these systems. This is a first step in checking whether these systems are realistic (a first validation). Indeed, in our present case, whenever the initial condition is positive the corresponding solution has to be positive as long as it exists. In our framework, such a property of solution can easily be proved. This is because, for system (4.47) or (4.48), Conditions (4.7) in Corollary 4.2 are trivially satisfied. For example, with 0 in place of x_1, the right-hand side of the first equation of (4.48) becomes sx_2. Since $s > 0$, we have $sx_2 \geq 0$ for all $x_2 \geq 0$, and so on.

It was shown by Murray [1] that the rectangular box

$$D = \{(x_1,x_2,x_3);\ 1 \leq x_1 \leq q^{-1};\ y_1 \leq x_2 \leq y_2;\ 1 \leq x_3 \leq q^{-1}\} \qquad (4.49)$$

has the property that every solution in the positive octant $x_i > 0$, $i = 1,2,3$, eventually enters the box D and no trajectory can leave D, provided that

$$y_1 = qf/(1 + q),\ y_2 = f/2q. \qquad (4.50)$$

In connection with this aspect, using Corollary 4.1 we can easily prove:

Proposition 4.6 *The necessary and sufficient conditions for the rectangular box* (4.49) *to be flow-invariant with respect to* (4.48) *are given by*

$$y_1 \leq qf/(1 + q),\ y_2 \geq f/2q. \qquad (4.50)'$$

Proof One applies Corollary 4.1 with $a_1 = a_3 = 1$, $a_2 = y_1$, $b_1 = b_3 = 1/q$ and $b_2 = y_2$. Since $s, f, w > 0$ and $0 < q < 1$, Conditions (B3)' in Lemma 4.1 are now equivalent to the following two inequalities:

$$fx_3 \geq (1 + x_1)y_1,\ fx_3 \leq (1 + x_1)y_2, \qquad (4.51)$$

where $1 \leq x_1 \leq 1/q$ and also $1 \leq x_3 \leq 1/q$. Obviously, (4.51) and (4.50)' are equivalent. This completes the proof.

Remark 4.4 From the point of view of invariance, our result here is more general than the above result of Murray. This is because (4.50)' gives all D of the form (4.49) which are invariant with respect to system (4.48). The determination of flow-invariant sets is also important in the study of the existence of oscillatory solutions of finite amplitude and of the limit cycle. (See R.J. Field and R.M. Noyes [1], S.P. Hasting and J.D. Murray [1], G. Bourceanu and G. Morosanu [1].) Moreover, the existence of a bounded flow-invariant set D with respect to an autonomous system (say (4.48)), implies the continuation to the whole $[0,+\infty[$ of all solutions starting in D. Next we point out the importance of this fact.

The following chemical scheme of P. Hanusse [1] exhibits limit cycle behaviour of two processes A and B with three intermediate species X, Y and Z:

$$A + X \underset{k_2}{\overset{k_1}{\rightleftarrows}} 2X, \quad X + Y \underset{k_2}{\overset{k_2}{\rightleftarrows}} 2Y,$$
$$Y + Z \rightleftarrows 2Z, \quad X + Z \underset{k_4^-}{\overset{k_4}{\rightleftarrows}} X + B. \tag{4.52}$$

Similarly to (4.47), the kinetic system for (4.49) is

$$\begin{aligned} x' &= k_1 ax - k_1^- x^2 - k_2 xy + k_2^- y^2 \\ y' &= k_2 xy - k_2^- y^2 - k_3 yz + k_3^- z^2 \\ z' &= k_3 yz - k_3^- z^2 - k_4 xz + k_4^- bx \end{aligned} \tag{4.53}$$

where all the constants k, a and b are positive. Obviously, for every initial condition $(x_0, y_0, z_0) \equiv u_0 \in R^3$ there is a unique solution $(x(t), y(t), z(t)) = u(t)$ with $u(0) = u_0$, $u : [0, t_{max}[\to R^3$ and t_{max} depending on u_0.

It may happen that $t_{max} < 0$. Indeed, in the case

$$\begin{aligned} x' &= 2x - y - x^2 - xy + y^2 \\ y' &= 2xy - 2\,yz - 2y^2 + 2z^2 \\ z' &= 3x - z + \frac{3}{2}\,yz - \frac{1}{2}\,z^2 - \frac{1}{2}\,xz \end{aligned} \tag{4.54}$$

the unique solution corresponding to $u_0 = (1,2,3)$ is as follows:

$$x(t) = 1/(1-t), \; y(t) = 2/(1-t), \; z(t) = 3/(1-t) \tag{4.54'}$$

with $t \in [0,1[$, that is $t_{max} = 1$. (The simplest example is: $x' = x^2$, $x(0) = 1$, with the solution $x(t) = (1-t)^{-1}$, $t \in [0,1[$.)

Now let us assume that $D \subset R^3$ is a bounded invariant set with respect to (4.53), and $u_0 \in D$. Then the corresponding solution u is defined on the whole $[0,+\infty[$, that is, $t_{max} = +\infty$. Indeed, we now have $u(t) \in D$ and therefore u is bounded on $[0,t_{max}[$. This implies $t_{max} = +\infty$ (see Proposition 2.3)).

P. Hanusse demonstrated numerically the existence of a stable limit cycle for (4.53), in the case

$$k_1 = k_2 = k_3 = k_4 = 1;\ k_1^- = k_2^- = k_3^- = 0{,}1;\ k_4^- = 0{,}05,\ a = b = 1. \quad (4.55)$$

Using Corollary 4.1, as in the proof of Proposition 4.6, it can be shown that the rectangular box

$$D_0 = \{(x,y,z);\ \frac{1}{3} \leqslant x \leqslant 10,\ \frac{1}{8} \leqslant y \leqslant 100;\ \frac{1}{20} \leqslant z \leqslant 1000\} \quad (4.56)$$

is invariant with respect to the system (4.53) (in the situation (4.55)). (See the proof of Theorem 4.7 below. A different proof can be found in Bourceanu and Morosanu [1].)

The final conclusion of this subsection is given by:

Theorem 4.7 (1) *The positive octant* R_+^3 *is a flow-invariant set with respect to each of the systems* (4.47), (4.48) *and* (4.53).

(2) *All the rectangular boxes of the form* (4.49) *which are invariant with respect to* (4.48) *are given by* (4.50)'.

(3) *Under Conditions* (4.55), *the rectangular box* D_0 *given by* (4.56) *is flow-invariant with respect to system* (4.53).

Proof Parts (1) and (2) have already been proved. We now prove (3). Under Conditions (4.55), system (4.53) becomes

$$\begin{aligned} x' &= x - \frac{1}{10}x^2 - xy + \frac{1}{10}y^2, \\ y' &= xy - \frac{1}{10}y^2 - yz + \frac{1}{10}z^2, \\ z' &= yz - \frac{1}{10}z^2 - xz + \frac{5}{100}x. \end{aligned} \quad (4.57)$$

To prove that D_0 (given by (4.56)) is flow-invariant with respect to (4.57), we apply Corollary 4.1 with $a_1 = \frac{1}{3}$, $b_1 = 10$, $a_2 = \frac{1}{8}$, $b_2 = 100$, $a_3 = \frac{1}{20}$, $b_3 = 1000$; $x_1 = x$, $x_2 = y$, $x_3 = z$. Clearly, the "tangential" conditions (B3)' in Lemma 4.1 yield

(I1) $\frac{1}{3} - \frac{1}{10}\frac{1}{9} - \frac{y}{3} + \frac{y^2}{10} \geqslant 0;$

(I2) $\frac{x}{8} - \frac{1}{10}\frac{1}{64} - \frac{z}{8} + \frac{z^2}{10} \geqslant 0;$

(I3) $\frac{y}{20} - \frac{1}{10}\frac{1}{400} - \frac{x}{20} + \frac{5x}{100} \geqslant 0;$

(I4) $-10y + \frac{y^2}{10} \leqslant 0;$

(I5) $100x - 1000 - 100z + \frac{z^2}{10} \leqslant 0;$

(I6) $1000y - 10^5 - 10^3x + \frac{5x}{100} \leqslant 0; \frac{1}{3} \leqslant x \leqslant 10, \frac{1}{8} \leqslant y \leqslant 100, \frac{1}{20} \leqslant z \leqslant 1000.$

Inequality (I1) is satisfied since

$$\min_{y \in R} \left\{ \frac{y^2}{10} - \frac{y}{3} + \frac{29}{90} \right\} = \frac{4}{10}.$$

To check (I2), it suffices to see that for $x = 1/3$ we have

$$\frac{1}{8}\frac{1}{3} \geqslant \frac{1}{8} \max \left\{ \frac{1}{640} + \frac{z}{8} - \frac{z^2}{10} \; ; \; z \in R \right\} = \frac{1}{8}\frac{13}{45} .$$

(I3) and (I4) are obviously true. For $x = 10$, (I5) reduces to $z(z-1000) \leqslant 0$, which is obvious for $1/20 \leqslant z \leqslant 1000$.

Finally, for $y = 100$ Inequality (I6) is trivially satisfied. The proof is complete.

Corollary 4.5 *Every solution* $u(t) = (x(t), y(t), z(t))$ *of* (4.57) *with* $u(t_o) \in D_o$ (*given by* (4.56)) *is defined on* $[t_o,+\infty[$ (*i.e.*, $u:[t_o,+\infty[\to D_o$, $\forall t_o \in R$).

Proof Since D_o is invariant with respect to (4.57), we have $u:[t_o,t_{max}[\to D_o$, that is, u is bounded on $[t_o,t_{max}[$. This implies $t_{max} = +\infty$ (by Proposition 2.3). □

§5. Flow invariance in Caratheodory type conditions

Throughout this section X is a finite dimensional space (e.g., $X = R^m$). We are concerned with the existence of D-valued solutions (to the right of t_o)

of Problem (4.1) in Caratheodory-type conditions on $A(t)x$. Let $\mathbb{A}$ be an open subset of X and let $D \subset \mathbb{A}$ be locally closed (see §3). The main result of this section is given by the following:

Theorem 5.1 *Suppose that the following conditions are fulfilled:*

(i) *For each* $t \in]a,b[$, *the function* $x \to A(t)x$ *is continuous from* D *into* X *and for each* $x \in D$, $t \to A(t)x$ *is Lebesgue integrable on each compact subset of* $]a,b[$ (*i.e., locally integrable on* $]a,b[$).

(ii) *For each compact* $K \subset D$ *there is a locally (Lebesgue) integrable function* $m:]a,b[\to R$ *such that* $\|A(t)x\| \leqslant m(t)$ *for all* $(t,x) \in]a,b[\times K$.

(iii) $\liminf_{h\downarrow 0} h^{-1} d(x + \int_t^{t+h} A(s)x\,ds;D) = 0, \quad \forall t \in]a,b[, x \in D.$

Let $t_0 \in]a,b[$, $x_0 \in D$, $r > 0$ *and* $T \in]t_0,b[$ *be such that* $D \cap B(x_0,r)$ *is closed and* $\int_{t_0}^{T} m(s)ds < r$, *where* m *is the function corresponding to* $K = D \cap B(x_0,r)$. *Then there is a continuous function* $u:[t_0,T] \to D$ *such that*

$$\|u(t) - u(s)\| \leqslant \int_s^t m(\tau)d\tau, \quad t_0 \leqslant s \leqslant t \leqslant T,$$

and $u'(t) = A(t)u(t)$, *a.e. on* $[t_0,T]$.

Proof Let t_0 and x_0 be as in the hypotheses of the theorem. Then for all sufficiently large positive integers n, we have $\int_{t_0}^{T} (m(s) + \frac{1}{n})ds \leqslant r$. We shall prove the existence of two sequences $\{t_i^n\}_{i=1}^{\infty}$ and $\{x_i^n\}_1^{\infty}$ with the properties:

$$t_0^n = t_0,\ x_0^n = x_0,\ x_i \in K$$

$$\|x_i^n - x_0\| \leqslant \int_{t_0}^{t_i^n} (m(s) + \tfrac{1}{n})ds, \quad i = 0,1,... \tag{5.1}$$

$$t_i^n < t_{i+1}^n \leqslant T \text{ and } \lim_{i\to\infty} t_i^n = T.$$

Since there is no danger of confusion we shall omit n as superscript for t_i and x_i. Given $t \in [t_0,T]$ and $x \in D$, define $\delta(t,x)$ by

$$\delta(t,x) = \sup \{h \in]0,1/n];\ d(x + \int_t^{t+h} A(s)x\,ds;D) \leq \frac{h}{2n}\}. \tag{5.2}$$

Inductively define $\{t_i^n\}$ and $\{x_i^n\}$ as follows: If $t_i^n = T$, set $t_{i+1} = t_i$ and $x_{i+1} = x_i$; if $t_i < T$, set $t_{i+1} = t_i + h_i$, with $h_i \in [\frac{1}{2}\delta(t_i,x_i), \delta(t_i,x_i)[$.

By the definition of $\delta(t_i,x_i)$ we have

$$d(x_i + \int_{t_i}^{t_i+h_i} A(s)x_i\,ds;D) \leq \frac{h_i}{2n}.$$

This shows that there is an element $x_{i+1} \in D$ such that

$$\|x_i + \int_{t_i}^{t_i+h_i} A(s)x_i\,ds - x_{i+1}\| \leq \frac{h_i}{n}.$$

Therefore it can be written in the form (see also (2.8))

$$x_{i+1} = x_i + \int_{t_i}^{t_i+h_i} A(s)x_i\,ds + h_iP_i,\ \|P_i\| \leq \frac{1}{n}. \tag{5.3}$$

Then, by induction hypothesis (5.1), (5.3) yields

$$\|x_{i+1} - x_0\| \leq \|x_i - x_0\| + \int_{t_i}^{t_{i+1}} m(s)\,ds + n^{-1}h_i$$

$$\leq \int_{t_0}^{t_i} (m(s) + \frac{1}{n})ds + \int_{t_i}^{t_{i+1}} (m(s) + \frac{1}{n})ds$$

$$= \int_{t_0}^{t_{i+1}} (m(s) + \frac{1}{n})ds,$$

hence $x_{i+1} \in K$.

Let us prove that $\lim_{i\to\infty} t_i = T$. Assume by contradiction that $\lim_{i\to\infty} t_i = L < T$. Then it follows that $\lim_{i\to\infty} h_i = 0$ and hence $\delta(t_i,x_i) \to 0$ as $i \to \infty$. Moreover, in view of (5.3) we have

$$\|x_{i+1} - x_i\| \leq \int_{t_i}^{t_{i+1}} (m(s) + \frac{1}{n})ds,\quad i = 0,1,\ldots,$$

which implies that $\lim_{i\to\infty} x_i = x^*$ exists. Choose $h^* \in]0,1/n[$ such that

$$L + h^* \leq T,\ d(x^* + \int_L^{L+h^*} A(s)x^*\,ds;D) \leq \frac{h^*}{4n}. \tag{5.4}$$

On the other hand, $\delta(t_i,x_i) < h^*$ for all $i \geq i_0$ (for a positive integer i_0 large enough). On the basis of the definition of $\delta(t_i,x_i)$ one has

$$d(x_i + \int_{t_i}^{t_i+h^*} A(s)x_i\,ds;D) > \frac{h^*}{2n},\quad \forall i \geq i_0. \tag{5.5}$$

Letting $i \to \infty$ in (5.5), one obtains

$$d(x^* + \int_L^{L+h^*} A(s)x^*\,ds;D) \geq \frac{h^*}{2n}$$

which contradicts (5.4).

We now define the functions $u_n:[t_0,T] \to D$ and $v_n:[t_0,T] \to X$ by

$$u_n(t) = \begin{cases} x_i & ,\ \text{if } t_i \leq t < t_{i+1} \\ \lim_{i\to\infty} x_i, & \text{if } t = T \end{cases} \tag{5.6}$$

$$v_n(t) = x_0 + \int_{t_0}^{t} A(s)u_n(s)ds,\ t \in [0,T]. \tag{5.7}$$

It is easy to prove that

$$\|v_n(t) - u_n(t) \leq \int_{t_i}^{t_{i+1}} m(s)ds + \frac{t_i-t_0}{n},\ t_i \leq t < t_{i+1}. \tag{5.8}$$

To do this, we first show that

$$\|v_n(t_i)-x_i\| \leq \frac{t_i-t_0}{n},\ i = 0,1,\ldots . \tag{5.9}$$

Indeed, for $i = 0$, (5.9) holds. Assume that (5.9) holds for all positive integers $j = 0,1,\ldots,i$. Then by (5.3) and (5.7) we have

$$\begin{aligned} v_n(t_{i+1}) - x_{i+1} &= x_0 + \int_{t_0}^{t_{i+1}} A(s)u_n(s)ds - x_i \\ &\quad - \int_{t_i}^{t_{i+1}} A(s)x_i ds - h_iP_i \\ &= v_n(t_i) - x_i - h_iP_i. \end{aligned}$$

Consequently, by induction hypothesis,

$$\|v_n(t_{i+1}) - x_{i+1}\| \leq \|v_n(t_i) - x_i\| + (t_{i+1} - t_i)/n$$

$$\leq (t_{i+1} - t_0)/n$$

and thus (5.9) holds true. Since $u_n(t) = x_i$ for $t_i \leq t < t_{i+1}$, we have

$$\|v_n(t) - u_n(t)\| \leq \|v_n(t) - v_n(t_i)\| + \|v_n(t_i) - x_i\|$$

$$\leq \int_{t_i}^{t_{i+1}} \|A(s)u_n(s)\|\, ds + \frac{t_i - t_0}{n}$$

which proves (5.8). Clearly, (5.8) yields $\lim_{n\to\infty} \|v_n(t) - u_n(t)\| = 0$, uniformly on $[t_0,T]$. On the other hand,

$$\|v_n(t)-v_n(s)\| \leq \left|\int_s^t m(u)du\right|, \quad t,s \in [t_0,T] \tag{5.10}$$

and

$$\|v_n(t) - x_0\| \leq \int_{t_0}^{T} m(s)ds \leq r.$$

By the Ascoli-Arzela lemma we may assume that $\lim_{n\to\infty} v_n(t) \equiv u(t)$ (uniformly on $[t_0,T]$). Inasmuch as $u_n(t) \in D \cap B(x_0,r)$ (which is closed), we see that $u(t) \in D \cap B(x_0,r)$ for all $t \in [t_0,T]$. Letting $n \to \infty$ in (5.7) (and using the Lebesgue theorem), one obtains

$$x_0 + \int_{t_0}^{t} A(s)u(s)ds,$$

which concludes the proof.

Theorem 5.1 implies the following result of flow invariance type.

<u>Corollary 5.1</u> *Let* $\mathbb{A}$ *be an open subset of* X. *Assume that Hypotheses* (i) *and* (ii) *of Theorem* 5.1 *are satisfied with* $\mathbb{A}$ *in place of* D. *Let* $D \subset \mathbb{A}$ *be a closed subset. If Hypothesis* (iii) *holds then at least a solution* u *to the problem* $u'(t) = A(t)u(t)$, *a.e. on* $[t_0,T]$, *starting in* D (*i.e.*, $u(t_0) = x_0 \in D$) *remains in* D *for all* $t \in [t_0,T]$, *for some* $T > t_0$ *sufficiently near to* t_o.

Remark 5.1 Condition (iii) of Theorem 5.1 is not necessary. However, it is not "too far" from necessity, in the following sense. Let u be a D-valued solution to $u' = A(t)u$, a.e. on $[t_0,T]$. Assume that u is differentiable at $t \in]t_0,T[$ and that t is a Lebesgue point of

$$\int_t^{t+h} A(s)u(t)ds$$

(i.e., $\lim_{h\downarrow 0} \frac{1}{h} \int_t^{t+h} A(s)u(t)ds = A(t)u(t)$). Then for $y = u(t) \in D$,

$$\lim_{h\downarrow 0} \frac{1}{h} d(y + \int_t^{t+h} A(s)yds;D) = 0. \tag{5.11}$$

Indeed,

$$h^{-1} d(y + \int_t^{t+h} A(s)yds;D) \leqslant \left\|\frac{u(t+h)-u(t)}{h} - \frac{1}{h} \int_t^{t+h} A(s)yds\right\|.$$

Since in the present case $u'(t) = A(t)u(t)$, the result (5.11) follows.

It is easy to see that (ii) implies

$$\liminf_{h\downarrow 0} h^{-1} d(x + hA(t)x;D) = 0 \tag{5.12}$$

a.e. on $]a,b[$, for each $x \in D$.

Remark 5.2 Condition (iii) is equivalent to: (iii)' For $x \in D$ and $t \in]a,b[$ there is $r:R_+ \to X$ with $r(h) \to 0$ as $h \downarrow 0$ and, in addition,

$$x + \int_t^{t+h} A(s)x\, ds + hr(h) \in D, \quad \forall h > 0.$$

Theorem 5.1 can be used in controllability theory. Let us sketch a very simple example in this direction, by considering the system

$$x' = Ax + u(t),\ x(t_0) = x_0 \tag{5.13}$$

where $A = (a_{ij})$, $i,j = 1,\dots,m$ with $a_{ij} \geqslant 0$ for $i \neq j$. Suppose that we are interested in getting positive solutions to (5.13). It follows from Theorem 5.1 that *iff* $u_i(t) \geqslant 0$, a.e. on $]a,b[$, $i = 1,\dots,m$, then the corresponding solution $x = (x_1,\dots,x_m)$ to (5.13) satisfies $x_i(t) \geqslant 0$, a.e. on $]a,b[$. Indeed, in this case $D = R_+^m$ (defined by (4.6)), and therefore (iii) is equivalent to

$$\sum_{\substack{j=1\\ j\neq i}}^{m} a_{ij}x_j + u_i(t) \geqslant 0, \text{ a.e. on }]a,b[\tag{5.14}$$

for all $x = (x_1,\ldots,x_m) \in R_+^m$ and $i = 1,\ldots,m$ (see Remark 5.2). Clearly, (5.14) holds *iff* $u_i(t) \geqslant 0$, a.e. on $]a,b[$ for $i = 1,\ldots,m$. Here u is supposed to be a.e. continuous (on $]a,b[$). If we are interested in bounded solutions: $a_i \leqslant x_i(t) \leqslant b_i$, $i = 1,\ldots,m$, then one proceeds as in Lemma 4.2 (see also the proof of Lemma 4.1).

§6. Notes and Remarks

The first counterexample to Peano's theorem in infinite dimensional Banach spaces was given by Dieudonné (1950) in c_o (see Dieudonné [2, p. 207]). We present here a slightly simplified version of this counterexample. As usual, c_o denotes the Banach space (with supremum norm) of all sequences $\{x_n\}$ which converge to zero as $n \to \infty$. Let us consider the function

$$f(x) = \{2\sqrt{|x_n|}\},\ x = \{x_n\} \in c_o.$$

Clearly $f:c_o \to c_o$ is continuous. However, the Cauchy problem

$$x'(t) = f(x(t)),\ x(0) = x_o,\ x_o = \{1/n^2\} \in c_o \tag{6.1}$$

admits no solutions on any nonvanishing interval $[-a,a]$, $a > 0$. Indeed, if Problem (6.1) were to admit a solution $x = \{x_n(t)\} \in c_o$ on an interval $[-a,a]$, then we would have

$$x_n'(t) = 2\sqrt{|x_n(t)|}\ ,\ x_n(0) = 1/n^2,\ -a \leqslant t \leqslant a,\ n = 1,2,\ldots\ . \tag{6.2}$$

Obviously, (6.2) leads to $x_n(t) = (t + \frac{1}{n})^2$, $0 \leqslant t \leqslant a$, $n = 1,\ldots$. We have reached the contradiction $0 = \lim_{n\to\infty} x_n(a) = a^2 > 0$. Consequently, (6.1) has no solution. In the original example of Dieudonne, $f(x) = \{\sqrt{|x_n|} + \frac{1}{n}\}$.

Other counterexamples to Peano's theorem in infinite dimensional spaces have been given by Yorke (1970), Godunov (1972), Lasota and Yorke (1973), in Hilbert spaces. However, the first general result in this direction was proved by Cellina (1972). Precisely, Cellina's result asserts that on every nonreflexive space X there exists a continuous function $F:R \times X \to X$ such that the corresponding Cauchy problem (1.1) admits no solution on any nonvanishing interval containing the origin. Moreover, Cellina [2, p. 1.9] has propounded the hypothesis that the nonexistence of solutions depends merely on the infinite dimensionality of the space (and not on other properties such as nonreflexivity). In other words, Cellina has stated without proof Theorem 1.1. This theorem was proved by Godunov (1975) and completed by Pianigiani [1], as indicated in Theorem 1.3. See also De Blasi and J. Myjak [1]. There are also results on the existence of weak solutions to Equation (1.1) on reflexive spaces.

The tangential condition (A4) in Section 2.1 has been used by the author and Vrabie [2]. There, a result similar to Theorem 2.1 is given. Precisely, in the paper above, we have supposed the continuity of $t \to B(x_0,r) \cap D(t)$ in Hausdorff mentric, in place of (A2). In the form presented here, Theorem 2.1 is due to Kenmochi and Takahashi [2]. An extension of this thoerem is Theorem 3.1 (in Chapter 5) due to the author [15]. For other types of generalizations, with $g(t, \|x-y\|)$ in place of $L(t)\,\|x-y\|^2$, we refer to S. Kato [1], Flett [1] and Vidossich [1,2]. Condition (A4) has also been considered by Yorke [1] in the finite dimensional case. Section 2 follows Kenmochi and Takahashi [2]. However, the idea to write u_ε and v_ε in the forms (2.18) and (2.22), respectively, is taken from the paper [2] by the author and Ursescu. This fact is frequently used in this book.

Theorem 3.2 is also due to Kenmochi and Takahashi [2] and improves a result of Martin [2]. Precisely, for global existence only, Martin has considered the additional condition: "$(t,x) \to A(t)x$ maps bounded subsets into bounded subsets". The first remarkable fact due to Kenmochi and Takahashi is the construction of Martin's sequence v_ε, without the dissipativity assumption on $A(t)$. A second consists of the method of proving the global existence of the solution to (2.1). Theorem 3.3, due to Martin (1973), is the fundamental result on the global existence of solutions in the continuous, autonomous case. In the nonautonomous case, with $D = X$, the result is given by Theorem 3.7, which has independently been proved by Lovelady & Martin (1972) and the author (1972), [1-2]. The second proof of this theorem is taken from the author's paper [2]. In the case $D = X$, Theorem 3.3, has also been proved by Martin (1970), who observed the remarkable fact that the continuity and dissipativity ensure the global existence of the solution to $u' = Au$.

Theorem 3.9 is taken from the author and Ursescu [1]. Propositions 3.2, 3.3 as well as Corollaries 3.1-3.3 are due to Martin [3], while Theorems 3.5 and 3.6 are due to Browder [2].

The tangential condition (3.9) goes back to Bouligand (1930) and Severi (1930) (see their papers [1], p. 32 and p. 99, respectively). If D is a submanifold, then (3.9) is equivalent to the fact that Ax is tangent to D at x, in the classical sense (see Chapter 4, Theorem 1.1, which is proved by Motreanu and the author [1]). In the theory of differential equations, Condition (3.9) has been used by Nagumo (1942) and, independently, by Brezis (1970). Yorke's paper has already been mentioned above. In the infinite dimensional case, Theorem 4.2 has been proved by Brezis (1970) and significantly extended by Martin (1973) (Theorem 3.3 in this chapter is such an extension). Other authors who have followed up Brezis' paper [1] include Crandall [1], Redheffer [1], Pavel and Ursescu [3], Motreanu and Pavel [1]. See also Abraham and Marsden [1] and Clarke [1].

The first major application of Theorem 4.2, given by Bourguignon and Brezis [1], was to the Euler equation, which describes the motion of an incompressible perfect fluid. Other applications of this theorem are given in Section 4, as well as in Chapter 3. The second proof of Theorem 4.2 is given by the author. Corollaries 4.1-4.3 were established by the author and Turinici [1]. The idea of considering the flow-invariance of time-dependent rectangular boxes given by (4.10) is due to Voicu [1]. The component-wise positive asymptotic stability (CWPAS) that we study in the Subsection 4.4 is a special case of the component-wise asymptotic stability defined by Voicu [1].

Theorem 4.6 is due to the author. A Fortran program to compute a solution of the inequality $Av < 0$ is due to O. Gherasim [1], who uses the algorithm in the proof of Theorem 4.6. The results in Subsections 4.3, 4.5 and 4.6 are proved by the author. The flow-invariance of the interior of the set given by (4.56) is proved by Bourceanu and Morosanu [1].

Theorem 5.1 is due to Ursescu. It was published (in 1977) in the author's book [9]. A similar result (with X - a Banach space) was independently obtained by Larrieu (1981). Recently, Ursescu [2] has extended his result from finite dimensional spaces to Fréchet spaces. Moreover, he proves (by using a result of Scorza Dragoni [1]) that the conclusion of Theorem 5.1 holds if and only if the tangential condition (iii) is fulfilled a.e. on

]a,b[. For other comments on Peano's theorem in Banach spaces (especially in compactness conditions), we refer to Deimling [1].

We end this section with some remarks which are useful in many important problems. We first give:

Lemma 6.1 *Let* X *be a Banach space and let* $f:D \subset X \to X$ *be locally Lipschitz* (§4, (B5)), *where* D *is open. Suppose that for an element* $x_0 \in D$, *there exist* $r > 0$ *and* $M > 0$, *such that* $B(x_0,r) \subset D$ *and* f *is bounded by* M *on* $B(x_0,r)$, *i.e.*

$$\|f(x)\| \leq M, \quad \forall \|x-x_0\| \leq r. \tag{6.3}$$

Let $T > 0$ *be such that* $TM < r$. *Then the Cauchy problem*

$$(E) \ x'(t) = f(x(t)), \ x(0) = x_0$$

has a unique solution $u:[0,T] \to D$, $\|u(t) - x_0\| \leq TM$, $\forall t \in [0,T]$.

Proof Since f is Lipschitz continuous in a neighbourhood $B(x_0,r_0)$ of x_0 with $r_0 \leq r$), there exists a unique solution $x^0:[0,T_0] \to D$ of the problem (E), with $0 < T_0 \leq T$ and $\|x^0(t) - x_0\| \leq TM$, $\forall t \in [0,T_0]$. Let t_m be the supremum of all such T_0 and let x_m be the solution of (E) on $[0,t_m[$. Clearly, $\|x_m(t) - x_0\| \leq TM$, $\forall t \in [0,t_m[$. Combining (6.3) and (E) we conclude that $\lim_{t \uparrow t_m} x_m(t) \equiv x^*$ exists and $\|x^*-x_0\| \leq TM < r$. It follows that x_m can be extended on $[0,t_m]$ as a solution of (E), which implies $t_m = T$, q.e.d. Another well-known result that we need is the following (for its proof see, e.g., Deimling [1, p.5]).

Lemma 6.2 *Let* D *be an open subset of the Banach space* X *and let* $f:D \to X$ *be a continuous function. Then for each* $\varepsilon > 0$, *there exists a locally Lipschitz function* $f_\varepsilon:D \to X$ *such that*

$$\|f_\varepsilon(x) - f(x)\| \leq \varepsilon, \quad \forall x \in D.$$

We now consider the following particular case of Theorem 2.1.

Theorem 6.1 *Let* D *be an open subset of* X *and* $B:D \to X$ *continuous and dissipative. Then, for each* $x_0 \in D$, *the problem*

(CP) $u' = Bu$, $u(0) = x_o$

has a unique local solution.

In the sequel we give a simple proof (essentially due to Brezis - private communication) of Theorem 6.1.

Let $B_n : D \to X$ be a locally Lipschitz function such that $\|B_n(x) - B(x)\| \leq 1/n$ for all $x \in D$ and $n = 1,2,\ldots$. Since B is continuous at x_o, there exist $r > 0$ and $M > 1$ with the property $\|Bx\| \leq M-1$, $\forall \|x - x_o\| \leq r$. Then

$$\|B_n(x)\| \leq M, \ \forall \|x - x_o\| \leq r, \ n = 1,2,\ldots .$$

On the basis of Lemma 6.1 there exists a unique solution u_n of the problem (with $TM < r$)

$$u_n'(t) = B_n(u_n(t)), \ u_n(0) = x_o, \ t \in [0,T], \ n = 1,2,\ldots .$$

We have

$$u_n' - u_m' = B_n(u_n) - B(u_n) + B(u_n) - B(u_m) + B(u_m) - B_m(u_m), \ t \in [0,T]$$

for all positive integers m and n. Since B is dissipative, with standard arguments (see the proof of (2.30)), one obtains

$$\|u_n(t) - u_m(t)\| \leq T(\frac{1}{n} + \frac{1}{m}), \ t \in [0,T]. \qquad (6.4)$$

It follows that $u(t) = \lim_{n\to\infty} u_n(t)$, $t \in [0,T]$, is the (unique) solution of (CP), on $[0,T]$. (Clearly, if B is merely continuous, then (6.4) fails. This approach can also be used in the theory of integral solutions (see Chapter 5, §6).

Moreover, let us observe that $\|u_n(t) - u_n(s)\| \leq M(t-s)$ for all $t,s \in [0,T]$. Consequently, if B is merely continuous and X is of *finite dimension*, then u_n is relatively compact in $C = C([0,T];X)$. We may assume (relabelling if necessary) that u_n is convergent (say, to u) in C. This u is a solution of (CP). In such a manner, we have given a new proof of Peano's theorem (i.e., a proof which uses neither fixed point theorems nor the classical method of polygonal lines).

We should mention that the tangential condition (3.10) (due to Y. Kobayashi [1]) plays a leading role in the abstract theory of nonlinear operators and semigroups, due to its unifying effect. Some comments in this direction, can be found in the author's papers [12, 14].

Finally, the following two remarks are important.

Remark 6.1 Let D be a closed subset of R^m and let $A:D \to R^m$ be continuous and satisfy the tangential condition (B3) (§4). Theorem 4.3 asserts that, for each $x_0 \in D$, at least a solution to

$$u' = Au, \ u(0) = x_0 \tag{6.5}$$

remains in D (to the right of 0), i.e., D is flow-invariant to the right, or, in other words, D is *forward invariant* for (6.5). In view of Tietze's theorem, we may assume that A is continuous even on R^m. Now, the following problem arises (when we have no uniqueness):

(Q) Are there solutions (to (6.5) starting in D, which do not lie in D?

It is easily seen that the answer is affirmative. Indeed, let us consider for simplicity the case $X = R^2$. Then System (4.5) becomes

$$x_1' = f_1(x_1,x_2); \quad x_2' = f_2(x_1,x_2) \tag{6.6}$$

with $f_i:R^2 \to R^2$-continuous, i = 1,2. Take two distinct solutions $x = (x_1(t),x_2(t))$, $\tilde{x} = (\tilde{x}_1(t),\tilde{x}_2(t))$, $t \in [0,T]$ with $x(0) = \tilde{x}(0) = x_0$ (see Figure 1). Set $D = \{x(t);\ t \in [0,T[\}$. Clearly, $f = (f_1,f_2)$ is tangent to D

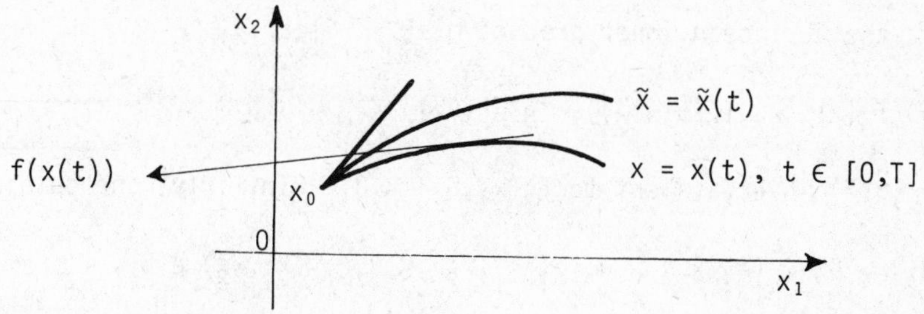

Figure 1

in the classical sense, hence (B4) holds (with f in place of A(t)). However, the solution $\tilde{x}$ which starts in $x_0 \in D$ does not lie in D on [0,T .

Remark 6.2 The invariance conditions (B3)' (Lemma 4.1) can be derived geometrically. Indeed, in this case D_0 is the rectangle

$$D_0 = \{x = (x_1, x_2);\ a_1 \leq x_1 \leq b_1,\ a_2 \leq x_2 \leq b_2\}.$$

Say a_1, $a_2 > 0$. Let $x = x(t)$ be a solution to (6.6) and let $P = P(x_1, b_2)$ on the boundary of D_0 (as in Figure 2). Denote by $\vec{n}$ the outward normal to D_0

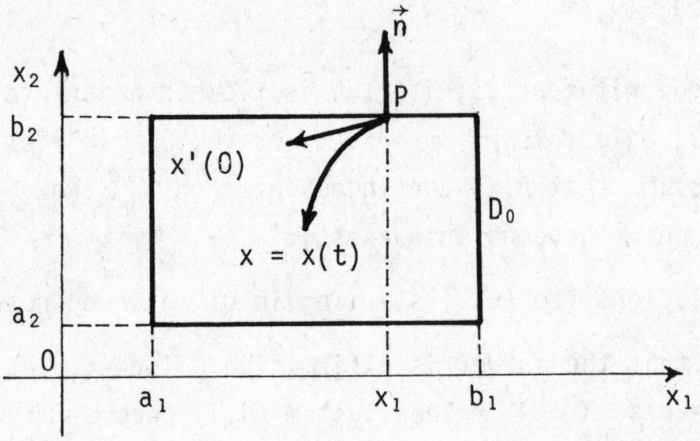

Figure 2

at P. Suppose that $x_1(0) = x_1$ and $x_2(0) = b_2$. For the flow-invariance of D_0 with respect to (6.6) we must have (necessarily and sufficient)

$$\langle x'(0), \vec{n}\rangle \leq 0 \tag{6.7}$$

where $\langle\ ,\ \rangle$ is the Euclidean inner product of R^2. Clearly,

$$x'(0) = (f_1(x_1, b_2), f_2(x_1, b_2)),\quad \vec{n} = (0,1).$$

Consequently, (6.7) is equivalent to $f_2(x_1, b_2) \leq 0$. Similarly, one obtains

$$f_2(x_1, a_2) \geq 0,\ f_1(a_1, x_2) \geq 0,\ f_1(b_1, x_2) \leq 0,\quad \forall (x_1, x_2) \in D_0.$$

This geometrical approach to flow invariance goes back to Wazewski (see Hartman [1]) and to Bony (see Crandall [1]). The same method has been used by Field and Noyes [1], Bourceanu and Morosanu [1], Murray [1], and by many others.

3 Flow invariance with respect to second-order differential equations and applications to flight space

§1. The main results on the flow invariance of a set with respect to a second-order differential equation

Throughout this chapter, A is a nonempty subset of a real Banach space X and $f:A \to X$ is a locally Lipschitz function. Let D be a nonempty subset of A.

Definition 1.1 *The nonempty set* D *is said to be closed in* A *if* $D = \bar{D} \cap A$, *where* $\bar{D}$ *is the closure of* D (*in* X). For the definition of the flow invariance of D with respect to the differential equation

$$u'(t) = f(u(t)),\ t = 0 \tag{1.1}$$

we refer to Chapter 2 (§4).

An obvious version of Theorem 4.2 in Chapter 2, is the following:

Theorem 1.1 *Suppose that the nonempty subset* D *is closed in* A. *Then* D *is a flow-invariant set with respect to* (1.1), *if and only if:*

$$\lim_{h \downarrow 0} \frac{1}{h} d(x + hf(x); D) = 0, \quad \forall x \in D. \tag{1.2}$$

Remark 1.1 If D is closed in X and f is globally Lipschitz on D, then the "tangential" condition (1.2) guarantees the existence of a D-valued solution u to (1.1) on R_+ (i.e., for each $x \in D$ there exists $u:R_+ \to D$ satisfying (1.1) on R_+ and $u(0) = x$. Such a solution is unique (see Theorem 3.3, Ch. 2).

We are now interested to define the notion of "flow-invariant set" for the autonomous second-order differential equation

$$u''(t) = f(u(t)),\ u(0) = x,\ u'(0) = y,\ t \geqslant 0, \tag{1.3}$$

For this purpose the following set M_D is needed:

$$M_D = \{(x,y) \in A \quad X;\ \lim_{h \downarrow 0} h^{-2}\, d(x+hy + \frac{h^2}{2} f(x);D) = 0\}. \tag{1.4}$$

Indeed, even in the next Theorem 1.2 we see that a necessary condition for the existence of a solution $u:[0,T[\to D$ to the problem (1.3) is that $(x,y) \in M_D$.

Theorem 1.2 *Denote by* $u:[0,T[\to A$ *a solution to* (1.3).

(i) *If* $u(t) \in D$, *for all* $t \in [0,T[$, *then* $(u(t),u'(t)) \in M_D$ *for all* $t \in [0,T[$.

(ii) *If* D *is closed in* A, *then* $u(t) \in D$ *if and only if* $(u(t),u'(t)) \in M_D$ *for all* $t \in [0,T[$.

The proofs of all results in this section are given in Section 3.

We are now prepared to give:

Definition 1.2 *The nonempty set* D *is said to be a flow-invariant set with respect to* (1.3) *if* M_D *is nonempty and if for every solution* $u:[0,T[\to A$ *to* (1.3) *with* $(u(0), u'(0)) \in M_D$, *we have* $u(t), u'(t)) \in M_D$ *for all* $t \in [0,T[$.

The relationship between this definition and Definition 4.1 in Chapter 2 is pointed out by:

Theorem 1.3 *Let* D *be closed in* A. *Then* D *is a flow-invariant set with respect to* (1.3) *if and only if* M_D *is nonempty and every solution* $u:[0,T[\to A$ *to* (1.3) *with* $(u(0), u'(0)) \in M_D$ *remains in* D *as long as it exists (i.e.,* $u(t) \in D, \ \forall t \in [0,T[$).

A simple consequence of this theorem is given by:

Corollary 1.1 *Let* D_j, $j \in J$ *be nonempty sets which are closed in* A (J *is a set of indexes*).

(1) *If each* D_j *is flow-invariant with respect to* (1.3) *and* $D = \bigcap_{j \in J} D_j \neq \emptyset$, *then*

$$M_D = \bigcap_{j \in J} M_{D_j} \tag{1.5}$$

(2) D *is flow-invariant for* (1.3) *iff* M_D *is nonempty.*

The main result on the flow-invariance of a set D with respect to the second-order equation (1.3) is the following:

Theorem 1.4 *Suppose that* M_D *is closed in* $A \times X$. *Then* D *is flow-invariant with respect to* (1.3) *if and only if* M_D *is nonempty and the following tangential condition holds:*

$$\lim_{h \downarrow 0} \frac{1}{h} d((x,y) + h(y,f(x)); M_D) = 0, \quad \forall (x,y) \in M_D. \tag{1.6}$$

A first simple application of the above result is the following:

<u>Theorem 1.5</u> *Let* S *be a closed linear subspace of* X *and* $D = A \cap S$. *Then*

(i) $M_D = (D \cap f^{-1}(S)) \quad S$.

(ii) *If in addition* $f(D) \subset S$, *then* $M_D = D \times S$ *and* D *is a flow-invariant set for* (1.3).

(Here, $f^{-1}(S) = \{z \in X;\ f(z) \in S\}$ and $f(D) = \{f(x);\ x \in D\}$.)

Next we are interested to examine (1.4) and (1.6) in significant particular cases.

Let Y be a real normed space (whose norm is also denoted by $\|\cdot\|$. We shall give some consequences of Theorem 1.4 in the case in which

$$D = D_g = \{x \in A,\ g(x) = 0\} \tag{1.7}$$

where $g: A \to Y$ is a function.

For this purpose we need some elements of Frechet differential calculus. Such elements can be found in Cartan ([1] Ch. I, §5). Let us recall here some basic tenets of Fréchet differentiability.

The function $g: A \to Y$ is said to be Fréchet differentiable at $x \in A$ if there is a linear continuous function (say, $\dot{g}(x)$) from X into Y such that

$$\lim_{\substack{y \to 0 \\ y \neq 0}} \frac{1}{y} \|g(x+y) - g(x) - \dot{g}(x)(y)\| = 0. \tag{1.8}$$

The function g is said to be differentiable on A if it is differentiable at every point $x \in A$.

Throughout this chapter the differentiability of a function is considered in the Fréchet sense, only.

By definition, $\dot{g}: A \to L(X,Y)$ (the space of all linear continuous operators from X into Y, endowed with the standard linear structure). Similarly, $\dot{g}$ is said to be differentiable at $x \in A$ if there is a linear continuous function (call it $\ddot{g}(x)$) from X into L(X,Y), such that

$$\lim_{\substack{y \to 0 \\ y \neq 0}} \frac{1}{\|y\|} \|\dot{g}(x+y) - \dot{g}(x) - \ddot{g}(x)(y)\| = 0 \tag{1.9}$$

where $\|z\|$ is the norm in $X(L(X,Y))$ if $z \in X$ (resp. $z \in L(X,Y)$). g is said to be twice differentiable on A if both g and $\dot{g}$ are differentiable at any $x \in A$.

Therefore, $\ddot{g}:A \to L(X,L(X,Y))$. Inductively one defines $\dddot{g}:A \to L(X,L(X,L(X,Y)))$ and so on. We shall also use the following consequence of the Taylor formula (when g is twice differentiable on $x \in A$):

$$\lim_{\substack{y\to 0\\ y\neq 0}} \frac{1}{\|y\|^2} \|g(x+y)-g(x)-\dot{g}(x)(y) - \frac{1}{2}\ddot{g}(x)(y)(y)\| = 0. \tag{1.10}$$

By $C^k(A,Y)$ we mean the set of all k-times differentiable functions $g:A\subset X\to Y$, with $g^{(k)}$ (the derivative of k-order) continuous on A. Here $g^{(1)} = \dot{g}$, $g^{(2)} = \ddot{g}$, $g^{(3)} = \dddot{g}$, for simplicity of writing.

In the sequel we shall suppose that $Y = R^n$ (the Euclidean n-space) and that the function $w:A \to R^n$ given by

$$w(x) = \dot{g}(x)(f(x)), \quad x \in A \tag{1.11}$$

is differentiable on A. For applications in mechanics the following three results are important.

<u>Theorem 1.6</u> *Suppose that the function* $u \to \dot{g}(x)(u)$ *from* X *into* R^n *is surjective (for each* $x \in D_g$*) and* g *is twice differentiable on* A. *Then* M_D *given by* (1.4) *has the following form:*

$$M_{D_g} = \{(x,y) \in A \times X; g(x) = 0, \dot{g}(x)(y) = 0, \ddot{g}(x)(y)(y)+\dot{g}(x)(f(x)) = 0\}. \tag{1.12}$$

Suppose in addition that g *is three times differentiable on* A, w *(given by* (1.11)) *is differentiable on* A, M_{D_g} *is nonempty and the function* $u \to (\dot{g}(x)(u), \ddot{g}(x)(y)(u))$ *from* X *into* $R^n \times R^n$ *is surjective, for each* $(x,y) \in M_{D_g}$. *Then* D_g *is a flow-invariant set for* (1.3) *iff*

$$\dddot{g}(x)(y)(y)(y)+2\ddot{g}(x)(f(x))(y)+\dot{w}(x)(y) = 0, \quad (x,y) \in M_{D_g}. \tag{1.13}$$

In the special case n = 1, the following result holds:

<u>Theorem 1.7</u> (1) *Let* $g:A \subset X \to R$ *be twice differentiable and* $w:A \to R$ *(giv*

by (1.11)) *satisfying the condition* $w(x) \neq 0$ *for all* $x \in D_g$. *Then* (1.12) *holds.*

(2) *Assume in addition that* g *is three times differentiable on* A, w *is differentiable on* A, *and* M_{D_g} *is nonempty. Then* D_g *is flow-invariant for* (1.3) *iff* (1.13) *holds.*

(3) *Moreover, for each* $x, \bar{y} \in X$ *with the properties*

$$g(x) = 0, \ \dot{g}(x)(\bar{y}) = 0, \ \dot{g}(x)(f(x))\ddot{g}(x)(\bar{y})(\bar{y}) < 0, \tag{1.14}$$

the pair $(x,y) \in M_{D_g}$, *where*

$$y = \sqrt{-\frac{\dot{g}(x)(f(x))}{\ddot{g}(x)(\bar{y})(\bar{y})}}\ \bar{y}. \tag{1.15}$$

In the case $X = R^2$, $Y = R$, the following notation is needed:

$$x = \begin{pmatrix} x_1 \\ x_2 \end{pmatrix} \in R^2, \ f(x) = \begin{pmatrix} f_1(x_1,x_2) \\ f^2(x_1,x_2) \end{pmatrix}, \ y = \begin{pmatrix} y_1 \\ y_2 \end{pmatrix}, \ y^2 = y_1^2 + y_2^2 = \|y\|^2 \tag{1.15'}$$

$$\begin{cases} g_i = g_i(x) = \dfrac{\partial g}{\partial x_i}(x), \ g_{ij} = g_{ij}(x) = \dfrac{\partial^2 g}{\partial x_i \partial x_j}(x), \\ \qquad\qquad g_{ijk} = g_{ijk}(x) = \dfrac{\partial^3 g}{\partial x_i \partial x_j \partial x_k}(x) \\ w_i = w_i(x) = \dfrac{\partial w}{\partial x_i}(x), \ i,j,k = 1,2. \end{cases} \tag{1.16}$$

In this case it is well known that

$$\begin{cases} \dot{g}(x) = \begin{pmatrix} g_1 \\ g_2 \end{pmatrix} \text{ (i.e., } \dot{g}(x) = \text{grad } g(x)), \ \ddot{g}(x) = \begin{pmatrix} g_{11} & g_{12} \\ g_{21} & g_{22} \end{pmatrix}, \\ \dot{g}(x)(y) = \langle g(x),y \rangle = g_1y_1+g_2y_2, \ \dot{g}(x)(\dot{g}(x)^{\perp}) = 0, \ \dot{g}(x)^{\perp} = \begin{pmatrix} g_2 \\ -g_1 \end{pmatrix} \end{cases} \tag{1.17}$$

where $\langle \cdot,\cdot \rangle$ stands for the inner product in R^2.

For the sake of simplicity, denote

$$a_1 = a_1(x) = g_1^2 + g_2^2 \text{ (i.e., } a_1(x) = \|\dot{g}(x)\|^2 = \|\dot{g}(x)^{\perp}\|^2)$$

$$a_2 = a_2(x) = g_{11}g_2^2 - 2g_{12}g_1g_2 + g_{22}g^2 \tag{1.18}$$

$$a_3 = a_3(x) = g_{111}g_2 - 3g_{112}g_1\, g_2^2 + 3g_{122}g_1^2g_2 - g_{222}g_1^3$$

where g_i^2 is the square of the number $g_i(x)$, $i = 1,2$.

Obviously,

$$\ddot{g}(x)(y)(y) = (y_1,y_2)\begin{pmatrix} g_{11} & g_{12} \\ g_{21} & g_{22} \end{pmatrix}\begin{pmatrix} y_1 \\ y_2 \end{pmatrix}$$
$$= g_{11}y^2 + 2g_{12}y_1y_2 + g_{22}y_2^2 \text{ (if } g_{12} = g_{21}) \tag{1.19}$$

$$\dddot{g}(x)(y)(y)(y) = \sum_{i,j,k=1}^{2} g_{ijk}y_iy_jy_k. \tag{1.20}$$

Therefore, with $y = \dot{g}(x)^{\perp}$, it follows that

$$a_2 = \ddot{g}(x)(\dot{g}(x)^{\perp})(\dot{g}(x)^{\perp}),\ a_3 = \dddot{g}(x)(\dot{g}(x)^{\perp})(\dot{g}(x)^{\perp})(\dot{g}(x)^{\perp}). \tag{1.21}$$

Denote by $c(x)$ the curvature of D_g at x. It is well known that

$$c(x) = \frac{a_2(x)}{(a_1(x))^{3/2}}\,. \tag{1.22}$$

From Theorem 1.7 we can derive a result with a unifying effect in the theory of flight space, namely:

<u>Theorem 1.8</u> *Assume that* $g:A \subset R^2 \to R$ *is twice differentiable and*

$$\dot{g}(x)(f(x))\ \ddot{g}(x)(y)(y) < 0 \tag{1.23}$$

for all (x,y) *with* $g(x) = 0$ *and* $\dot{g}(x)(y) = 0$, $y \neq 0$. *Then* M_{D_g} *given by* (1.12) *has the form*

$$M_{D_g} = \{(x,y) \in A \times R^2;\ g(x) = 0,\ g_1y_1+g_2y_2 = 0,\ y^2 = -\frac{\dot{g}(x)(f(x))}{c(x)\ \|\dot{g}(x)\|}\} \tag{1.24}$$

and for each $x \in D_g$ *there is* $y \in R^2$ *such that* $(x,y) \in M_{D_g}$. *Assume in addition that* g *is three times differentiable on* A *and* $w : A \subset R^2 \to R$ (*given by* (1.11)) *is differentiable on* A. *Then* D_g *is flow-invariant for* (1.3) *iff*

$$-\dot{g}(x)(f(x))\frac{a_3(x)}{a_2(x)} + 2(g_{11}g_2 - g_{12}g_1)f^1(x) + 2(g_{21}g_2 - g_{22}g_1)f^2(x) + w_1g_2 - w_2g_1 = 0 \qquad (1.25)$$

for all x *with* g(x) = 0.

Remark 1.2 If $f : A \subset R^2 \to R^2$ is differentiable on A, the invariance condition (1.13) obviously becomes

$$\dddot{g}(x)(y)(y)(y) + 3\ddot{g}(x)(f(x))(y) + \dot{g}(x)(\dot{f}(x)(y)) = 0, \qquad (1.26)$$

since

$$\dot{w}(x)(y) = \ddot{g}(x)(f(x))(y) + \dot{g}(x)(\dot{f}(x)(y)).$$

In this case (1.25) can be rewritten in the form

$$-(g_if^1 + g_2f^2)\frac{a_3(x)}{a_1(x)} + 3(g_{11}g_2 - g_{12}g_1)f^1 + 3(g_{21}g_2 - g_{22}g_1)f^2 + g_1g_2\left(\frac{\partial f^1}{\partial x_1} - \frac{\partial f^2}{\partial x_2}\right) + g_2^2\frac{\partial f^2}{\partial x_1} - g_1^2\frac{\partial f^1}{\partial x_2} = 0 \qquad (1.27)$$

for all x with g(x) = 0.

Indeed, this partial differential equation in f^i is a direct consequence of (1.25), in which w_i (i = 1,2) is replaced by expressions (1.29) below.

First of all, in this case (1.11) is the following function:

$$w(x) = \langle \dot{g}(x), f(x) \rangle = g_1f^1 + g_2f^2. \qquad (1.28)$$

Therefore

$$\dot{w}_1 = \dot{w}_1(x) = \frac{\partial w}{\partial x_i} = g_{1i}f^1 + g_{2i}f^2 + g_1\frac{\partial f^1}{\partial x_i} + g_2\frac{\partial f^2}{\partial x_i}, \; i = 1,2 \qquad (1.29)$$

$$\dot{w}(x)(\dot{g}(x)^{\perp}) = \langle \dot{w}(x), \dot{g}(x)^{\perp} \rangle = w_1g_2 - w_2g_1. \qquad (1.30)$$

§2. Tangent vectors to a subset of a Banach space. An inverse problem in Taylor formula

For the proof of the results stated in Section 1 it is necessary to develop a mathematical apparatus on tangent sets. We shall use the notation of the previous section.

Let us consider the conditions

$$\lim_{h\downarrow 0} \frac{1}{h}\, d(x + hy;D) = 0, \tag{2.1}$$

$$\lim_{h\downarrow 0} \frac{1}{h^2}\, d(x + hy + \frac{h^2}{2} z;D) = 0 \tag{2.2}$$

where D is a nonempty subset of the real Banach space X and $x,y,z \in X$.

In view of the nonexpansivity of the function $x \to d(x;D)$ (see (2.3)', Chapter 2), we have

$$\frac{1}{h}\, |d(x + hy + \frac{h^2}{2} z;D) - d(x + hy;D)| \le \frac{h}{2}\, \|z\|,$$

which shows that (2.2) implies (2.1). The set of all y satisfying (2.1) is said to be a "tangent set" (cone) to D at x. In some particular cases (e.g., when D is a smooth set) the tangent set is just the tangent space to D at x (in the classical sense). On this subject we refer to Chapter 4, Section 1.

Let E be a nonempty subset of real normed space Y, $A \subset X$ an open subset of X and $g:A \to Y$ a differentiable function at $x \in A$.

We now introduce the condition

$$\lim_{h\downarrow 0} \frac{1}{h}\, d(g(x) + h\dot{g}(x)(y);E) = 0 \tag{2.3}$$

and, if g is twice differentiable at $x \in A$,

$$\lim_{h\downarrow 0} \frac{1}{h^2}\, d(g(x) + h\dot{g}(x)(y) + \frac{h^2}{2} (\ddot{g}(x)(y)(y) + \dot{g}(x)(z));E) = 0. \tag{2.4}$$

The key of the proof of the results from Section 1 is given by the next two theorems.

Theorem 2.1 *If* $D = g^{-1}(E)$ *and* g *is differentiable at* $x \in A$ *then* (2.1) *implies* (2.3). *If in addition to the above hypothesis we assume that* g *is continuous on* A, $u \to \dot{g}(x)(u)$ *from* X *into* Y *is surjective and* Y *is finite dimensional, then*

(2.3) *implies* (2.1) (*therefore in this case* (2.1) *and* (2.3) *are equivalent*).

Theorem 2.2 *If* $D = g^{-1}(E)$ *and* g *is twice differentiable at* $x \in A$, *then* (2.2) *implies* (2.4). *If, in addition,* $u \to \dot{g}(x)(u)$ *from* X *into* Y *is surjective and* Y *is finite dimensional, then* (2.2) *is equivalent to* (2.4).

For the proof of these theorems, the following lemmas are useful:

Lemma 2.1 *Condition* (2.1) *is equivalent to each of the statements* (2.5), (2.6) *below.*

For every $\varepsilon > 0$, *there is* $\delta = \delta(\varepsilon) > 0$, *such that for each* $h \in]0,\delta[$ *there exists* $u = u(\varepsilon,h) \in X$ *satisfying*

$$x + h(y+u) \in D, \quad \|u\| < \varepsilon. \tag{2.5}$$

For each $h > 0$ *there is* $u(h) \in X$ *such that* $u(h) \to 0$ *as* $h \downarrow 0$ *and*

$$x + h(y + u(h)) \in D. \tag{2.6}$$

Lemma 2.2 *Each of the statements* (2.7) *and* (2.8) *is equivalent to* (2.2), *where for every* $\varepsilon > 0$, *there is* $\delta > 0$ *such that, for each* $h \in]0,\delta[$, *there is* $u = u(\varepsilon,h) \in X$ *with the properties*

$$\|u\| < \varepsilon, \; x + hy + \frac{h^2}{2}(z + u) \in D; \tag{2.7}$$

for each $h > 0$, *there is* $u(h) \in X$ *such that* $u(h) \to 0$ *as* $h \downarrow 0$ *and*

$$x + hy + \frac{h^2}{2}(z + u(h)) \in D. \tag{2.8}$$

In the Cartesian product $X \times X$, the last part of Lemma 2.1 becomes:

Lemma 2.3 *Let* M *be a nonempty subset of* $X \times X$. *For each* $(x,y) \in M$ *and* $(z,w) \in X \times X$, *the following conditions* (i) *and* (ii) *are equivalent:*

(i) $\lim_{h \downarrow 0} \frac{1}{h} d((x,y) + h(z,w);M) = 0.$

(ii) *For each* $h > 0$, *there exist* $r_j(h) \in X$ $(j = 1,2)$ *with* $r_j(h) \to 0$ *as* $h \downarrow 0$, *such that*

$$(x + h(z + r_1(h)), y + h(w + r_2(h))) \in M.$$

The proof of these lemmas is elementary and is indicated in Ch. 2, (3.9)', so omit it here. A second set of results establishes some consequences of the relations (1.8) and (1.9).

Let us first consider (1.8). If $g:A \subset X \to Y$ is differentiable at $x \in A$ then from (1.8) it follows that

$$\lim_{h \downarrow 0} \frac{1}{h}(g(x+hy) - g(x) - h\dot{g}(x)(y)) = 0 \tag{2.9}$$

uniformly with respect to y in bounded subsets of X. With y + u in place of y, (2.9) yields

$$\lim_{\substack{h \downarrow 0 \\ u \to 0}} \frac{1}{h}(g(x+h(y+u)) - g(x) - h\dot{g}(x)(y)) = 0. \tag{2.10}$$

The lemma below states a reciprocal (in a certain sense) relation to (2.10), or an inverse problem in Taylor formula:

Lemma 2.4 *Assume that* g *is continuous on* A, *differentiable at* $x \in A$, $u \to \dot{g}(x)(u)$ *from* X *into* Y *is surjective and* Y *is finite dimensional. Then for every* $\varepsilon > 0$ *there is* $\delta > 0$ *such that for each* $h \in]0,\delta[$ *and* $v \in Y$ *with* $\|v\| < \delta$, *there is* $u \in X$ *with the properties:*

$$\|u\| < \varepsilon,\ x+h(y+u) \in A,\ g(x+h(y+u)) = g(x)+h(\dot{g}(x)(y)+v). \tag{2.11}$$

In the proof of Lemma 2.4, the following simple result is needed:

Lemma 2.5 *Let* $L:X \to Y$ *be a linear surjective operator. If* Y *is finite dimensional, then there is a linear continuous operator* $T:Y \to X$ *such that*

$$L(T(y)) = y, \textit{ for all } y \in Y. \tag{2.12}$$

Proof Let $e_1,\ldots,e_n$ be a basis for Y. Choose x_i such that $L(x_i) = e_i$, $i = 1,2,\ldots,n$. If y is arbitrary in Y, there exists $a_i \in R$, $i = 1,\ldots,n$ such that $y = \sum_{i=1}^{n} a_i e_i$. With $T:Y \to X$ given by $T(y) = \sum_{i=1}^{n} a_i x_i$, the assertion of the lemma is proved.

Proof of Lemma 2.4 Since $\dot{g}(x)$ is surjective, by Lemma 2.5 (with $L = \dot{g}(x)$) there is a linear continuous operator $T:Y \to X$, such that

$$\dot{g}(x)(T(w)) = w, \text{ for all } w \in Y. \tag{2.13}$$

Let $\varepsilon > 0$. The continuity of T at 0 implies the existence of $r > 0$ such that

$$\|T(w)\| < \varepsilon, \text{ for all } w \in B(r) = \{w \in Y; \ \|w\| \leqslant r\}. \tag{2.14}$$

According to (2.10), there is $\delta = \delta(\varepsilon) > 0$, such that

$$x + h(y + T(w)) \in A$$

$$\frac{1}{h} \|g(x+h(y+T(w))) - g(x) - h\dot{g}(x)(y+T(w))\| < \frac{r}{2} \tag{2.15}$$

for all $h \in \,]0,\delta[$ and $w \in B(r)$.

We may assume (without loss of generality) that $\delta \leqslant r/2$. Let us show that this δ satisfies the condition required by our lemma. Take an arbitrary $h \in \,]0,\delta[$ and $v \in Y$ with $\|v\| < \delta$ and denote by $F:B(r) \to Y$ the function

$$F(w) = \frac{1}{h} [(-g(x+h(y+T(w))) + g(x)+h\dot{g}(x)(y+T(w)))] + v. \tag{2.16}$$

By the linearity of $\dot{g}(x)$ and (2.13), (2.16) yields

$$g(x+h(y+T(w)) = g(x)+h(\dot{g}(x)(y)+v)+h(w-F(w)). \tag{2.17}$$

In view of (2.15) and of $\delta \leqslant r/2$, $\|v\| \leqslant \delta \leqslant r/2$, we have $\|F(w)\| \leqslant r$ for all $w \in B(r)$, i.e., $F:B(r) \to B(r)$. Since F is continuous on $B(r)$, by the Brower fixed point theorem there is an element $w \in B(r)$ such that $F(w) = w$. With this w and $u = T(w)$, (2.14), (2.15) and (2.17) show that the requirements of the lemma are satisfied. The proof is complete.

We now assume that g is twice differentiable at $x \in A$. Replacing in (1.10) y by hy (y by $hy + \frac{h^2}{2} z$) and taking into account $\ddot{g}(x) \in L(X,L(X,Y))$, we easily get (2.18) (resp. (2.19)) below:

$$\lim_{h \downarrow 0} \frac{2}{h^2} [g(x+hy) - g(x)-h\dot{g}(x)(y) - \frac{h^2}{2} \ddot{g}(x)(y)(y)] = 0 \tag{2.18}$$

uniformly with respect to y in bounded subsets of X,

$$\lim_{h \downarrow 0} \frac{2}{h^2} [g(x+hy + \frac{h^2}{2} z)-g(x)-h\dot{g}(x)(y) - \frac{h^2}{2} (\ddot{g}(x)(y)(y)+\dot{g}(x)(z))] = 0 \tag{2.19}$$

uniformly with respect to (y,z) in bounded subsets of $X \times X$.

With $z + u$ in place of z, (2.19) yields

$$\lim_{\substack{h\downarrow 0\\ u\to 0}} \frac{2}{h^2}\left[g(x+hy+\frac{h^2}{2}(z+u))-g(x)-h\dot{g}(x)(y)-\frac{h^2}{2}(\ddot{g}(x)(y)(y)+\dot{g}(x)(z))\right]=0. \tag{2.20}$$

The lemma below establishes a reciprocal (in a certain sense) relation to (2.20).

Lemma 2.5 *Assume that* g *is continuous on* A, *twice differentiable at* $x \in A$, $u \to \dot{g}(x)(u)$ *from* X *into* Y *is surjective and* Y *is finite dimensional. Then for every* $\varepsilon > 0$, *there is* $\delta > 0$ *such that for each* $h \in]0,\delta[$ *and* $v \in Y$ *with* $\|v\| < \delta$, *there is* $u \in X$ *with the properties*

$$\begin{cases} \|u\| < \varepsilon,\ x + hy + \frac{h^2}{2}(z+u) \in A, \\ g(x + hy + \frac{h^2}{2}(z+u)) = g(x) + h\dot{g}(x)(y) + \frac{h^2}{2}(\ddot{g}(x)(y)(y) + \dot{g}(x)(z)+v). \end{cases} \tag{2.21}$$

Proof Let T be as in the proof of the previous lemma. For an arbitrary $\varepsilon > 0$, let $r > 0$ be such that (2.14) is true. Since A is open and (2.19) holds (with $z + T(w)$ instead of z) it follows that there is $\delta > 0$, with the properties

$$x + hy + \frac{h^2}{2}(z+u) \in A \tag{2.22}$$

and

$$\frac{2}{h^2}\|g(x+hy+\frac{h^2}{2}(z+T(w)))-g(x)-h\dot{g}(x)(y)-\frac{h^2}{2}(\ddot{g}(x)(y)(y) + \dot{g}(x)(z+T(w)))\| < \frac{r}{2} \tag{2.23}$$

for all $h \in]0,\delta[$ and $w \in B(r)$ (see (2.14)).

Obviously, we may assume that $\delta \leq r/2$. Let us show that this δ is a suitable one. Indeed, let $h \in]0,\delta[$ and $y \in Y$ with $\|v\| < \delta$. Define the function $F:B(r) \to Y$ by

$$F(w) = \frac{2}{h^2}\left[-g(x + hy + \frac{h^2}{2}(z + T(w))) + g(x) + h\dot{g}(x)(y) + \frac{h^2}{2}(\ddot{g}(x)(y)(y) + \dot{g}(x)(z + T(w)))\right] + v. \tag{2.24}$$

The linearity of $\dot{g}(x)$ and (2.24) imply

$$g(x + hy + \frac{h^2}{2}(z+T(w))) = g(x) + h\dot{g}(x)(y) + \frac{h^2}{2}(\ddot{g}(x)(y)(y)$$

$$+ \dot{g}(x)(z) + v + w-F(w)). \qquad (2.25)$$

Clearly, (2.23) and (2.24) yield $\|F(w)\| \leq r$ for all $w \in B(r)$, that is, $F:B(r) \to B(r)$. Since F is continuous on $B(r)$, there is $w \in B(r)$ such that $F(w) = w$. With this w and $u = T(w)$, (2.14), (2.22) and (2.25) conclude the proof.

We can now proceed to the proof of Theorems 2.1 and 2.2.

Proof of Theorem 2.1 Let us assume that g is differentiable at $x \in A$ and that (2.1) holds. In order to get (2.3) we shall use Lemma 2.1. By this lemma, for each $h > 0$ there is $u(h) \in X$ such that $u(h) \to 0$ as $h \downarrow 0$ and $x + h(y+u(h)) \in D = g^{-1}(E)$. This means that

$$g(x+h(y+u(h))) \in E, \quad \forall h > 0$$

and therefore

$$\frac{1}{h} d(g(x)+h\dot{g}(x)(y);E) \leq \frac{1}{h} \|g(x)+h\dot{g}(x)(y)-g(x+h(y+u(h)))\|$$

which, according to (2.10)), implies (2.3).

Let us now suppose that g is continuous on A, $\dot{g}(x):X \to Y$ is surjective, Y is finite dimensional and that (2.3) holds. Then by (2.6), for each $h > 0$, there is $r(h) \in Y$ with $r(h) \to 0$ as $h \downarrow 0$, such that

$$g(x) + h(\dot{g}(x)(y) + r(h)) \in E. \qquad (2.26)$$

For an arbitrary $\varepsilon > 0$, let $\delta = \delta(\varepsilon) > 0$ be a number with the property given by Lemma 2.4. Since $r(h) \to 0$, as $h \downarrow 0$, there is $\delta_1 = \delta_1(\varepsilon) \in]0,\delta[$ such that

$$\|r(h)\| < \delta, \quad \forall h \in]0,\delta_1[. \qquad (2.27)$$

According to Lemma 2.4, for each h with $0 < h < \delta_1 (\delta_1 < \delta)$ there is $u = u(h) \in X$ with the properties

$$\|u\| < \varepsilon,\ x+h(y+u) \in A,\ g(x+h(y+u)) = g(x)+h(\dot{g}(x)(y) + r(h)). \quad (2.28)$$

Clearly, (2.28) shows (taking into account (2.26)) that $x+h(y+u) \in g^{-1}(E) = D$. Consequently, in view of Lemma 2.1, (2.1) is proved.

Proof of Theorem 2.2 The proof of this theorem is similar to that of Theorem 2.1, as follows:

Let $D = g^{-1}(E)$, $g:A \to Y$-twice differentiable at $x \in A$ and assume that (2.2) holds.

By Lemma 2.2, there is $u(h) \in X$ with the property (2.8), which gives

$$g(x+hy + \frac{h^2}{2}(z+u(h))) \in E, \quad \forall h > 0.$$

Consequently we have

$$\frac{1}{h^2} d(g(x) + h\dot{g}(x)(y) + \frac{h^2}{2}(\ddot{g}(x)(y)(y) + \dot{g}(x)(z));E)$$
$$\leq \frac{1}{h^2} \|g(x) + h\dot{g}(x)(y) + \frac{h^2}{2}(\ddot{g}(x)(y)(y)$$
$$+ \dot{g}(x)(z)) - g(x+hy + \frac{h^2}{2}(z+u(h))\|.$$

Passing to the limit and taking into account (2.20), one obtains (2.4).

We now suppose that $\dot{g}(x):X \to Y$ is surjective, Y is finite dimensional and (2.4) holds. By (2.8) there is $r(h) \in Y$ such that $r(h) \to 0$ as $h \downarrow 0$ and, in addition,

$$g(x)+h\dot{g}(x)(y) + \frac{h^2}{2}(\ddot{g}(x)(y)(y) + \dot{g}(x)(z) + r(h)) \in E, \quad \forall h > 0. \quad (2.29)$$

For $\varepsilon > 0$, let $\delta > 0$ be a number with the property given by Lemma 2.4 and $\delta_1 \in]0,\delta[$ satisfying (2.27). Since $0 < \delta_1 < \delta$, according to Lemma 2.5, for each $h \in]0,\delta_1[$ there is $u \in X$ with the properties

$$\begin{cases} \|u\| < \varepsilon,\ x + hy + \frac{h^2}{2}(z+u) \in A \\ g(x+hy + \frac{h^2}{2}(z+u)) = g(x) + h\dot{g}(x)(y) + \frac{h^2}{2}(\ddot{g}(x)(y)(y)+\dot{g}(x)(z)+r(h)). \end{cases} \quad (2.30)$$

On the basis of (2.29), (2.30) yields

$$x + hy + \frac{h^2}{2}(z + u) \in g^{-1}(E) = D \quad (2.31)$$

and thus (2.2) follows. The proof is complete.

From Theorems 2.1 and 2.2 we easily get the results below:

Corollary 2.1 *Suppose that* $g:A \subset X \to Y$ *is differentiable at* $x \in A$ *continuous on* A, $\dot{g}(x):X \to Y$ *is surjective and* Y *is finite dimensional. Then Conditions* (2.32) *and* (2.33) *below are equivalent:*

$$\lim_{h\downarrow 0} \frac{1}{h} d(x + hy; D_g) = 0, \; y \in X \tag{2.32}$$

$$g(x) = 0, \; \dot{g}(x)(y) = 0 \tag{2.33}$$

where D_g *is given by* (1.7).

Corollary 2.2 *In addition to the hypotheses of Corollary* 2.1, *we assume that* g *is twice differentiable at* $x \in A$. *Then Conditions* (2.34) *and* (2.35) *below are equivalent:*

$$\lim_{h\downarrow 0} \frac{1}{h^2} d(x + hy + \frac{h^2}{2} z; D_g) = 0, \; y,z \in X \tag{2.34}$$

$$g(x) = 0, \; \dot{g}(x)(y) = 0, \; \ddot{g}(x)(y)(y) + \dot{g}(x)(z) = 0. \tag{2.35}$$

Proof of Corollaries 2.1 and 2.2. Let us denote also by 0 the null element of Y and let us set $E = \{0\}$. Then $D_g = g^{-1}(E)$. Accordingly, by Theorems 2.1 and 2.2, Conditions (2.32) and (2.34) are respectively equivalent to

$$\lim_{h\downarrow 0} \frac{1}{h} d(g(x) + h\dot{g}(x)(y); \{0\}) = 0$$

$$\lim_{h\downarrow 0} \frac{1}{h^2} d(g(x) + h\dot{g}(x)(y) + \frac{h^2}{2} (\ddot{g}(x)(y)(y) + \dot{g}(x)(z)); \{0\}) = 0$$

which completes the proof.

In the study of uniform motion we have to consider the case

$$g(x) = \frac{1}{2} (\|x\|^2 - r^2), \; r > 0. \tag{2.36}$$

Consequently,

$$D_g = S(r) \equiv \{x \in X; \; \|x\| = r\}. \tag{2.37}$$

We already know that in this case (2.32) is equivalent to $\|x\| = r$ and $\langle x,y\rangle_+ = 0$ (Lemma 3.2, Chapter 2).

Finally, let X be a real Hilbert space H of inner product $\langle\cdot,\cdot\rangle$ and norm $\|\cdot\|$. Then with g given by (2.36) we have

$$\dot{g}(x)(y) = \langle x,y\rangle, \ddot{g}(x)(y)(z) = \langle y,z\rangle, \quad \forall y,z \in H. \tag{2.38}$$

In addition, according to (2.9) in Chapter 2 one has

$$\langle y,x\rangle = \|x\| \ \langle y,x\rangle_+, \quad \forall x,y \in H. \tag{2.39}$$

Consequently, Corollaries 2.1 and 2.2 become:

Proposition 2.1 (1) *The condition* $\lim_{h\downarrow 0} h^{-1} d(x+hy;S(r)) = 0$ *is equivalent to*

$$\|x\| = r, \langle x,y\rangle = 0, y \in H.$$

(2) *For each* $x,y,z \in H$ *the condition*

$$\lim_{h\downarrow 0} h^{-2} d(x + hy + \frac{h^2}{2} z; S(r)) = 0$$

is equivalent to

$$\|x\| = r, \langle x,y\rangle = 0, \ \|y\|^2 + \langle x,z\rangle = 0. \tag{2.40}$$

Remark 2.1 The formulae given by (2.11) and (2.21) can be regarded as "inverse Taylor formulae".

§3. Proof of the results in Section 1

Proof of Theorem 1.2 Let $u:[0,T[\to A$ be a solution of (1.4). It is known that if $u''(t)$ exists then

$$\lim_{h\downarrow 0} \|u(t+h)-u(t)-hu'(t) - \frac{h^2}{2} u''(t)\| /h^2 = 0 \tag{3.1}$$

where the derivatives are taken in the strong sense (see Cartan [1], Ch. 1). Therefore

$$\lim_{h\downarrow 0} \|u(t+h)-u(t)-hu'(t) - \frac{h^2}{2} f(u(t))\| /h^2 = 0. \tag{3.2}$$

If $u(t) \in D$ for all $t \in [0,T[$, then for each $t \in [0,T[$, $u(t+h) \in D$ for all $h \in [0,T-t[$. Accordingly,

$$\frac{1}{h^2} d(u(t) + hu'(t) + \frac{h^2}{2} f(u(t));D) \leq \frac{1}{h^2} \|u(t) + hu'(t) + \frac{h^2}{2} f(u(t)) - u(t+h)\| \tag{3.3}$$

for all $h \in]0,T-t[$ and $t \in [0,T[$. Combining (3.2) and (3.3), it follows that $(u(t),u'(t)) \in M_D$. We now suppose that D is closed in A (in the sense of Definition 1.1) and $(u(t),u'(t)) \in M_D$ (given by (1.5)). This means (by Lemma 2.2) that there is $r(h) \in X$ with $r(h) \to 0$ as $h \downarrow 0$, such that

$$u(t) + hu'(t) + \frac{h^2}{2} (f(u(t)) + r(h)) \in D = \bar{D} \cap A, \quad \forall h > 0. \tag{3.4}$$

Since $u(t) \in A$, (3.4) implies $u(t) \in \bar{D} \cap A = D$.

Remark 3.1 With the same proof is follows that the projection $pr_1(M_D)$ of $M_D \subset A \times X$ on the first factor space of $X \times X$ satisfies $pr_1(M_D) \subset D \cap A$. Therefore the case $D = \bar{D} \cap A$ yields $pr_1(M_D) \subset D$. Theorem 1.3 is a consequence of Theorem 1.2 and Definition 1.2.

Proof of Corollary 1.1 The inclusion $M_D \subset \bigcap_{j \in J} M_{D_j}$ is a direct consequence of (1.4) and Lemma 2.2. For the converse inclusion take $(x,y) \in \bigcap_{j \in J} M_{D_j}$. Inasmuch as f is locally Lipschitz, the solution $u:[0,T[\to A$ of (1.3) is uniquely determined by $(x,y) \in M_{D_j}$. Since $(u(0),u'(0)) \in M_{D_j}$ it follows by Theorem 1.3 that $u(t) \in D_j$ for all $t \in [0,T[$ and $j \in J$, that is, $u(t) \in D$ for $t \in [0,t[$. This implies $(u(t),u'(t)) \in M_D$ and therefore $(x,y) \in M_D$. The proof of (2) is similar and is left to the reader.

Proof of Theorem 1.4 It is easily seen that D is a flow-invariant set with respect to (1.3) if and only if M_D is a flow-invariant set with respect to the system

$$\begin{aligned} u'(t) &= v(t) \\ v'(t) &= f(u(t)), \; t \geq 0. \end{aligned} \tag{3.5}$$

Inasmuch as f is locally Lipschitz from A into X, it follows that the function $(x,y) \to (y,f(x))$ from $A \times X$ into $X \times X$ is also locally Lipschitz. In view of Theorem 1.1 it is now clear that M_D is flow-invariant with respect to (3.5) if and only if (1.6) holds (i.e., for each $(x,y) \in M_D$ the vector $(y,f(x))$ is "tangent" to M_D at (x,y)).

<u>Proof of Theorem 1.5</u> On the basis of Lemma 2.2, if $(x,y) \in M_D$ then for each $h > 0$ there is $u(h) \in X$ such that $u(h) \to 0$ as $h \downarrow 0$ and, in addition,

$$x + hy + \frac{h^2}{2}(f(x) + u(h)) \in D = A \cap S \subset S, \quad \forall h > 0.$$

Since S is closed, (3.6) implies $x \in S$ so $x \in A \cap S = D$. Further, because S is a linear space, (3.6) now implies

$$y + \frac{h}{2}(f(x) + u(h)) \in S, \quad \forall h > 0. \tag{3.7}$$

Similarly, (3.7) gives $y \in S$ and then $f(x) \in S$. Therefore $x \in f^{-1}(S)$ and $(x,y) \in (D \cap f^{-1}(S)) \times S$.

Conversely, let $(x,y) \in (D \cap f^{-1}(S)) \times S$. Consequently, for each $h > 0$ one has $v_h = x + hy + \frac{h^2}{2} f(x) \in S$.

Since A is open, there is $\delta > 0$ such that $v_h \in A$ (hence $v_h \in A \cap S$), $\forall h \in]0,\delta[$. According to the definition of M_D (see (1.4)), it now trivially follows that $(x,y) \in M_D$ and Part (i) of the theorem is proved. If we assume that $f(D) \subset S$ (i.e., $D \subset f^{-1}(S)$) then obviously, (i) becomes $M_D = D \times S$. Therefore M_D is a nonempty set closed in $A \times X$. To get the last assertion of (ii), we apply Theorem 1.4.

Let $(x,y) \in M_D$ (which means $x \in A \cap S$ and $y \in S$). Then, for every $h > 0$, we have

$$(x,y) + h(y,f(x)) \in S \times S$$

and there is $\delta > 0$ such that $x + hy \in A$, $\forall h \in]0,\delta[$. Therefore

$$(x,y) + h(y,f(x)) \in (A \cap S) \times S = D \quad S = M_D, \ \forall h \in]0,\delta[$$

which trivially implies (1.6). The theorem is proved.

<u>Proof of Theorem 1.6</u> The form (1.12) of M_D given by (1.4) follows from Corollary 2.2. To prove the second part of the theorem, observe that M_{D_g} can

be written in a form similar to (1.7), namely

$$M_{D_g} = \{(x,y) \in A \times X;\ k(x,y) = 0\} \equiv M_{D_k} \tag{3.8}$$

with $k: A \times X \to R^{3n}$ given by

$$k(x,y) = (g(x), \dot{g}(x)(y), \ddot{g}(x)(y)(y) + w(x))$$

where w is defined by (1.11). Clearly, for each $(x,y) \in A \times X$ and $u,v \in X$ we have

$$\dot{k}(x,y)(u,v) = (\dot{g}(x)(u), \ddot{g}(x)(y)(u) + \dot{g}(x)(v), \dddot{g}(x)(u)(y)(y) + 2\ddot{g}(x)(y)(v) + \dot{w}(x)(u)).$$

We now prove that the function $\dot{k}(x,y): X \times X \to R^{3n}$ is surjective (for each $(x,y) \in M_{D_g}$).

To do this, let $y_i \in R^n$, $i = 1,2,3$. We have to prove that there is $(u,v) \in X \times Y$ such that $\dot{k}(x,y)(u,v) = (y_1,y_2,y_3)$, i.e.

$$\begin{aligned} &\dot{g}(x)(u) = y_1 \\ &\ddot{g}(x)(y)(u) + \dot{g}(x)(v) = y_2 \\ &\dddot{g}(x)(u)(y)(y) + 2\ddot{g}(x)(y)(v) + \dot{w}(x)(u) = y_3. \end{aligned} \tag{3.9}$$

The first hypothesis of the theorem is the surjectivity of $\dot{g}(x): X \to R^n$. Therefore there is $u \in X$ verifying the first equation of (3.9). The existence of $v \in X$ verifying the other two equations of (3.9) is a direct consequence of surjectivity of

$$u \to (\dot{g}(x)(u), \ddot{g}(x)(y)(u)).$$

By Corollary 2.1, with $k(M_{D_k})$ instead of g (resp. D_g), we conclude that (1.6) holds iff $\dot{k}(x,y)(y,f(x)) = 0$ holds (which leads us to (1.13)).

<u>Proof of Theorem 1.7</u> In this case (1.12) is a direct consequence of Theorem 1.6. Indeed, since $\dot{g}(x): X \to R$ and $\dot{g}(x)(f(x)) \neq 0$ on D_g (according to one of the hypotheses), $\dot{g}(x)$ is not the null functional and consequently is surjective.

For the second part of the theorem, it suffices to check that in this case the linear function $u \to (\dot{g}(x)(u), \ddot{g}(x)(y)(u))$ from X into R^2 is surjective (for each $(x,y) \in M_{D_g}$) or, equivalently, that the linear functions

$$u \to \dot{g}(x)(u), \quad u \to \ddot{g}(x)(y)(u) \tag{3.10}$$

from X to R are linear independent.

To this aim, let us consider the linear combination,

$$r\dot{g}(x)(u) + s\ddot{g}(x)(y)(u) = 0, \ r,s \in R, \ u \in X. \tag{3.11}$$

The hypothesis $w(x) \neq 0$ for all $x \in D_g$ implies (on the basis of (1.12) that $\ddot{g}(x)(y)(y) \neq 0$ for all $(x,y) \in M_{D_g}$. Therefore, with $u = y$, (3.11) gives $s = 0$ (since $\dot{g}(x)(y) = 0$). Furthermore, for $u = f(x)$, (3.11) gives $r = 0$ and hence, by Theorem 1.6, the flow-invariance of D_g is equivalent to (1.13). Finally, the third part of the theorem is obvious.

Proof of Theorem 1.8 Let $(x,y) \in M_{D_g}$ given by (1.12). Clearly, with $\bar{y} = \dot{g}(x)^{\perp}$ given by (1.17), (1.23) is satisfied. In this case, $\dot{g}(x)(y) = 0$ and $\dot{g}(x)(\bar{y}) = 0$ show that the vectors y and $\bar{y}$ are parallel, therefore there is $a = a(x) \in R$ such that $y = a\bar{y}$. Since $\ddot{g}(x)(y)(y) + \dot{g}(x)(f(x)) = 0$, it follows that

$$a^2\ddot{g}(x)(\bar{y})(\bar{y}) + \dot{g}(x)(f(x)) = 0. \tag{3.12}$$

Having in mind the notation of (1.15) - (1.22) and

$$||y||^2 = a^2 ||\bar{y}||^2 = a^2 ||\dot{g}(x)^{\perp}||^2 = a^2 a_1, \ \ddot{g}(x)(\bar{y})(\bar{y}) = a_2,$$

it is clear that (3.12) (and (1.22)) implies $(x,y) \in M_{D_g}$ given by (1.24). Note that Hypothesis (1.23) implies $w(x) \neq 0$ and (for $y = \dot{g}(x)^{\perp}$) $a_2 \neq 0$.

Actually we have proved that, for $x \in D_g$, there is $y \in X$, namely,

$$y = a\bar{y} = -\sqrt{\frac{\dot{g}(x)(f(x))}{a_2}}\ \dot{g}(x)^{\perp} \tag{3.13}$$

such that $(x,y) \in M_{D_g}$ and conversely (i.e., if $(x,y) \in M_{D_g}$ then y is given by (3.13)). Finally, by replacing y given by (3.13) in (1.13) and taking into account (1.30), one obtains (1.25).

§4. Some Applications in flight mechanics

The purpose of this section is to point out that Theorem 1.4 and its consequences contain some fundamental results of dynamics. Moreover, our results here (e.g., Theorem 4.2, Corollary 4.1, Theorem 4.4), which are obtained via the "flow-invariance technique", show that some basic results on the motion within a set D_g in R^2 or R^3 remain valid in any real Hilbert (or Banach) space.

4.1 The properties of a flow-invariant set

The function $f:A \to X$ defining the equation of "motion" (1.3) is regarded as a field of force on A. This means that with each vector position $x \in A$ is associated the vector force $f(x)$.

On the basis of Definition 1.2 and Theorem 1.3, the notion

$$\text{"}D_g \text{ is flow-invariant with respect to (1.3)"} \tag{4.1}$$

can be restated in terms of dynamics as follows:

> *a "mass particle projected" from a point* $x \in D_g$ *with a speed* $y \in X$ *such that* $(x,y) \in M_{D_g}$ *given by* (1.12) *describes, under the action of the force field* f, *an orbit which lies in* D_g. (4.2)

Let us consider again the function $g:A \subset X \to R$. Denote by N_x or $N(\dot{g}(x))$ the null space of $\dot{g}(x):X \to R$, that is,

$$N_x = \{y \in X;\ \dot{g}(x)(y) = 0\}. \tag{4.3}$$

In terms of geometry, N_x is the tangent space to D_g at x.

Definition 4.1 (1) *The function* g *from* $A \subset X$ *into* R *is said to be "smooth" if:*

(a) g *is three times (Fréchet) differentiable on* A.

(b) *For each* $x \in D_g$, $\dot{g}(x):X \to R$ *is surjective (i.e., there is* $u \in X$, *such that* $\dot{g}(x)(u) \neq 0$).

(c) *For each* $x \in D_g$, $\ddot{g}(x)(y)(y) \neq 0$, $\forall y \in N_x$, $y \neq 0$.

(2) g *is said to be "completely smooth", if it is smooth and if, for each* $x \in D_g$, *the functional* $F(x)$ *defined by*

$$F(x)(y) = \begin{cases} \dfrac{\dddot{g}(x)(y)(y)(y)}{\ddot{g}(x)(y)(y)}, & \text{if } y \in N_x,\ y \neq 0 \\ 0, & \text{if } y = 0 \end{cases} \tag{4.4}$$

is a linear continuous functional from N_x *into* R.

<u>Definition 4.2</u> *A field of force* f *on* A *is said to be "g-smooth" if it is locally Lipschitz, the function* $w:A \to R$ *(given by (1.11)) is differentiable on* A *and, for each* $x \in D_g$,

$$\dot{g}(x)(f(x))\, \ddot{g}(x)(y)(y) < 0, \quad \forall y \in N_x,\ y \neq 0. \tag{4.5}$$

<u>Theorem 4.1</u> *Let* A *be an open subset of the real Banach space* X *and let* $g:A \to R$ *be a smooth function.*

(1) *If there is a g-smooth force field* f *on* A *such that a "mass" particle "projected" from any point* $x \in D_g$ *with a "speed"* y *such that* $(x,y) \in M_{D_g}$ *"describes" (under the action of* f*) an "orbit" which lies in* D_g, *then* g *is completely smooth.*

(2) *If* g *is completely smooth and* f *is g-smooth, then* D_g *is a flow-invariant set for (1.3) iff, for each* $x \in D_g$, *there is* $a(x) \in R$ *with the property*

$$-w(x)F(x) + 2\ddot{g}(x)(f(x)) + \dot{w}(x) = a(x)\dot{g}(x), \quad x \in D_g. \tag{4.6}$$

<u>Proof</u> (1) Let g be a smooth function from A into R. If there is a g-smooth force field f on A with the property that D_g is a flow-invariant set for (1.3) then, in view of Theorem 1.7, (1.13) holds. Take $x \in D_g$. If $y \neq 0$ is an arbitrary element of N_x (given by (4.3)) and

$$\bar{y} = \left(- \frac{w(x)}{\ddot{g}(x)(y)(y)} \right)^{1/2} \quad y = \bar{a}y, \quad \bar{a} = \left(- \frac{w(x)}{\ddot{g}(x)(y)(y)} \right)^{1/2}, \tag{4.7}$$

then obviously $(x,\bar{y}) \in M_{D_g}$ (given by (1.12)).

Replacing $(x,\bar{y})$ in (1.13) and dividing by $\bar{a}$, it follows that

$$-w(x)F(x)(y) + 2\ddot{g}(x)(f(x))(y) + \dot{w}(x)(y) = 0 \tag{4.8}$$

for all $y \in N_x$. By definition of differentiability, the functions $y \to \dot{w}(x)(y)$, $y \to \ddot{g}(x)(f(x))(y)$ are linear continuous functionals from X into R. Therefore (4.8) implies that F(x) given by (4.4) is linear continuous from N_x into R, hence g is completely smooth.

(2) This is a consequence of (4.8). Indeed, let $L_i: X \to R$, $i = 1,2$ be two linear functionals (different from the null functional). Set

$$N(L_i) = \{y \in X,\ L_i(y) = 0\},\ i = 1,2.$$

It is well known that if $N(L_2) \subset N(L_1)$ then there is $a \in R$, $a \neq 0$, such that $L_1 = aL_2$ (which implies $N(L_1) = N(L_2)$). Since $N_x = N(\dot{g}(x))$, (4.6) follows from (4.8) with

$$L_1 = -w(x)F(x) + 2\ddot{g}(x)(f(x)) + \dot{w}(x),\quad L_2 = \dot{g}(x).$$

Remark 4.1 If g is smooth and f is g-smooth then M_{D_g} is nonempty. Moreover, for each $x \in D_g$ there is $y \in X$ (e.g., y given by (4.7)) such that $(x,y) \in M_{D_g}$.

We now give some examples of completely smooth functions and g-smooth force field. Let us consider the function

$$g(x) = \frac{1}{2}\left(\| x - \bar{a}\|^2 - (\langle \bar{b}, x - \bar{a}\rangle + d)^2\right),\ x \in R^n,\ n \geqslant 2, \tag{4.9}$$

where $\bar{a}, \bar{b} \in R^n$ and $d \in R$, $d \neq 0$, $\langle\ ,\ \rangle$ is the inner product of R^n. In the case $n = 2$, $C = D_g = \{x \in R^2 : g(x) = 0\}$ is a conic with $\bar{a}$ as one of the foci (i.e., circle, ellipse, hyperbola and parabola).

Proposition 4.1 *The function* $g: R^n \to R$ $(n = 2,3)$ *given by* (4.9) *is a completely smooth function and we have*

$$\ddot{g}(x)(y)(y) = y^2(1 - \cos^2[x - \bar{a}, y]) > 0 \tag{4.10}$$

for all $x,y \in R^n$ *with* $y \neq 0$, $g(x) = 0$, $\dot{g}(x)(y) = 0$. *Moreover, the Newtonian field on* $A \supset D_g$ *is* g-*smooth.*

Here $[x - \bar{a}, y]$ denotes the angle determined by the vectors y and $x - \bar{a}$

(which joins $x \in D_g$ with the focus $\bar{a}$).

Proof Property (a) appearing in Definition 4.1 is obviously satisfied in this case. (b) is also satisfied, since the gradient $\dot{g}(x)$ of a conic is different from the null vector. Of course, (4.10) is also a well-known property. However, we prove it here (since some of the formulae involved in its proof are needed in 4.2 - 4.4).

Clearly,

$$\dot{g}(x)(y) = \langle x-\bar{a},y\rangle - (\langle \bar{b},x-\bar{a}\rangle + d)\, \langle \bar{b},y\rangle. \tag{4.11}$$

The condition $g(x) = 0$ means that

$$\|x - \bar{a}\|^2 = (\langle \bar{b},x - \bar{a}\rangle + d)^2, \tag{4.12}$$

which implies $x \neq \bar{a}$ (since $d \neq 0$). Now, $\dot{g}(x)(y) = 0$ and $g(x) = 0$ yield

$$\langle \bar{b},y\rangle^2 = \frac{\langle x-\bar{a},y\rangle^2}{\|\bar{x}-a\|^2} = \|y\|^2 \cos^2 [x-\bar{a},y]. \tag{4.13}$$

On the other hand,

$$\ddot{g}(x)(y)(z) = \langle z,y\rangle - \langle \bar{b},z\rangle \langle \bar{b},y\rangle, \forall y,z \in R^n. \tag{4.14}$$

Hence $\dddot{g}(x) = 0$ and, in addition,

$$\ddot{g}(x)(y)(y) = y^{2} - \langle \bar{b},y\rangle^2. \tag{4.14)'}$$

Combining (4.13) and (4.14)', one obtains (4.10) (since $[x-\bar{a},y]$ is neither 0 nor 180°). In this case, $F(x)$ defined by (4.4) is the trivial functional. In 4.5 we shall see that the Newtonian field (4.73) is g-smooth.

Proposition 4.2 (1) *Any smooth function* $g:R^2 \to R$ *is completely smooth.* (2) *Any smooth polynomial* $g:R^n \to R$ *of (at most) second degree is completely smooth* (*e.g.*, (4.9)).

Proof (1) In this case $N_x = N(\dot{g}(x))$ given by (4.3) is a one-dimensional subspace of R^2. Therefore there is $r = r(x) \neq 0$ $(r \in R)$ such that, for any $y = \binom{y_1}{y_2} \in N(\dot{g}(x))$, we have $y_1 = ry_2$.

On the basis of (1.19) and (1.20), it follows that there is a constant $b = b(x) \in R$ such that

$$F(x)(y) = \begin{cases} b(x)y_2, & \text{if } y \in N_x,\ y \neq 0 \\ 0, & \text{if } y = 0 \end{cases}$$

hence $y \to F(x)(y)$ is a linear functional on R^2. The proof of the second part is obvious since in this case $\dddot{g}(x)(y)(y)(y) = 0$, $y \in R^n$.

Remark 4.2 It would be interesting to give an example of "completely smooth" function in R^n ($n > 3$) other than the smooth polynomials of (at most) second degree.

Proposition 4.3 *Let* $g:A \subset X \to R$ *be three times differentiable on* A.

(1) *If there is a* g*-smooth force field on* A, *then* g *is necessarily a* g*-smooth function.*

(2) *If* g *is smooth, then, for each* $x \in D_g$, *the linear continuous function* $\binom{u}{s} \to L_x\binom{u}{s}$ *from* $X \times R$ *into* $L(X,R) \times R$ *defined by*

$$(*)\ L_x\binom{u}{s} = \begin{pmatrix} \ddot{g}(x)(u)+s\dot{g}(x) \\ \dot{g}(x)(u) \end{pmatrix} \equiv \begin{pmatrix} \ddot{g}(x) & \dot{g}(x) \\ \dot{g}(x) & 0 \end{pmatrix} \binom{u}{s},\ u \in X,\ s \in R$$

is one-to-one.

(3) *Let* $X = R^n$ *and* $g:A \subset R^n \to R$ *be a smooth function. Then, for each* $x \in D_g$, L_x *is a bijection (i.e., one-to-one and surjective).*

(4) *For* $n = 2$, *the curvature* $c(x)$ *of* D_g *is different from zero.*

Proof (1) This part is a direct consequence of Definitions 4.1 and 4.2.

(2) Let $u \in X$ and $s \in R$ be such that $L_x\binom{u}{s} = 0$, i.e.,

$$\ddot{g}(x)(u) + s\dot{g}(x) = 0,\ 0 \in L(X,R),\ \dot{g}(x)(u) = 0 \text{ (i.e., } u \in N_x).$$

Then $u = 0$ and therefore also $s = 0$. Indeed, if $u \neq 0$, then $0 = \ddot{g}(x)(u)(u) + s\dot{g}(x)(u) = \ddot{g}(x)(u)(u)$ and $u \in N_x$ contradict Condition (c) of Definition 4.1.

(3) In this case L_x can be identified with the $(n+1) \times (n+1)$ matrix:

$$L_x^{n+1} = \begin{pmatrix} g_{11} \cdots g_{1n} & g_1 \\ g_{n1} \cdots g_{nn} & g_n \\ g_1 \cdots g_n & 0 \end{pmatrix}, \tag{4.14)"}$$

since $\dot{g}(x)$, $\ddot{g}(x)$, $L(R^n,R)$ are identified with gradient, Hessian matrix and R^n respectively.

The fact that L_x^{n+1} is one-to-one is equivalent to $\det(L_x^{n+1}) \neq 0$ (hence, to the surjectivity of L_x^{n+1}).

(4) If $n = 2$ it is easy to check that $- a_2(x) = \det\,(L_x^3) \neq 0$ which implies (in view of (1.22) and of $a_1(x) \neq 0$), that $c(x) \neq 0$. Actually, the fact that $a_1(x) \neq 0$ ($a_2(x) \neq 0$) follows directly from Hypotheses (b) and (c) (with $y = \dot{g}(x)^{\perp}$) of Definition 4.1, respectively (see the notation of (1.15) - (1.21).

4.2 Generalized Bonnet's theorem

As a first application of our results we give an extension of Bonnet's theorem (see, e.g., Whittaker [1], p. 95). Given $g:A \subset X \to R$, set

$$K_g = \left\{ \begin{array}{l} f:A \subset X \quad X;\ f \text{ is "g-smooth"} \\ \text{and (4.1) (or equivalently, (4.2)) holds} \end{array} \right\}. \tag{4.15}$$

For the statements of the results below we shall use Definitions 4.1 and 4.2 and the notation of (1.7) and (1.11).

Remark 4.3 If g is completely smooth, then, in view of Theorem 4.1, it follows that

$$K_g = \{f:A \to X;\ f \text{ is g-smooth and (4.6) holds}\}. \tag{4.16}$$

The null function 0 does not satisfy (4.5), hence $0 \in K_g$.

Theorem 4.2 *Let* X *be a real Banach space,* $A \subset X$ *an open subset and* $g:A \to R$ *a completely smooth function. Then* K_g *is a convex cone (which does not contain the null function). More precisely, if* $f_i \in K_g$ *and* $b_i \geqslant 0$, $i = 1,\ldots,m$, *with* $\sum_{i=1}^{m} b_i^2 > 0$, *then* f *given by*

$$f = \sum_{i=1}^{m} b_i f_i \tag{4.17}$$

also belongs to K_g. *Moreover, if* $x \in D_g$ *and* $v = v(x)$ *is such that* $(x,v) \in M_{D_g}$ *given by* (1.12) *with* f *defined by* (4.17), *then for each* $i = 1,\ldots,m$, *there is* v_i *having the properties*

$$(x,v_i) \in M^i_{D_g}, \quad v^2 = \sum_{i=1}^{m} b_i v_i^2. \tag{4.18}$$

Here we have denoted by $M^i_{D_g}$ the subset M_{D_g} corresponding to f_i, i.e.,

$$M^i_{D_g} = \{(x,y) \in A \times X; g(x) = 0, \dot{g}(x)(y) = 0,$$
$$\ddot{g}(x)(y)(y) + \dot{g}(x)(f_i(x)) = 0\}. \tag{4.19}$$

Proof of Theorem 4.2 Since f_i are g-smooth, then obviously f given by (4.17) is also g-smooth. The fact that $f_i \in K_g$, given by (4.15), means also that for each $x \in D_g$, there is $a_i = a_i(x) \in R$ such that

$$-w_i(x)F(x) + 2\ddot{g}(x)(f_i(x)) + \dot{w}_i(x) = a_i g(x), \; x \in D_g \tag{4.20}$$

where (similarly to (1.11))

$$w_i(x) = \dot{g}(x)(f_i(x)), \; x \in A \; (\text{hence } \sum_{i=1}^{m} b_i w_i(x) = w(x)). \tag{4.21}$$

Multiplying (4.20) by b_i and then summing over 1 to m, we get (4.6) with f given by (4.16) and $a = \sum_{i=1}^{m} b_i a_i$.

To prove the last assertion of the theorem, let $x \in D_g$ and $v \in X$ be such that $(x,v) \in M_{D_g}$. The existence of such a v is shown in Remark 4.1. Hence

$$g(x) = 0, \; \dot{g}(x)(v) = 0, \; \ddot{g}(x)(v)(v) + \dot{g}(x)(f(x)) = 0 \tag{4.22}$$

with f given by (4.17) (therefore $w(x) = \dot{g}(x)(f(x))$).

It is easy to check that v_i given by

$$v_i = \left(\frac{w_i(x)}{w(x)}\right)^{1/2} v \tag{4.23}$$

has the property that $(x,v_i) \in M^i_{D_g}$. Note that $f \in K_g$ implies $w(x) \neq 0$ (according to (4.5)) and then $v \neq 0$ (by (4.22)).

By (4.23) we have $b_i v_i^2 = (b_i w_i(x)/w(x))\, v^2$, which yields (4.18) and the proof is complete.

A classical theorem of Bonnet (restated here as Corollary 4.1) is a consequence of Theorem 4.2.

Corollary 4.1 *Let* $X = R^2$, $g:A \subset R^2 \to R$ *a completely smooth function. If the orbit* D_g *can be described in each* g-*smooth field of force* f_i, $i = 1,\ldots,m$, *the velocity of any point* P *of the orbit being* v_i, *then the same orbit can be described in the field of force*

$$f = \sum_{i=1}^{m} b_i f_i \quad (b_i > 0, \ \sum_{i=1}^{m} b_i^2 \neq 0),$$

the velocity v *of* P *(in the field of force* f*) being*

$$v^2 = \sum_{i=1} b_i v_i^2. \tag{4.24}$$

Proof In view of Theorem 4.2, the only fact we have to prove is (4.24). Let v_i be as in the statement of the corollary. In our framework this means (see, e.g., Theorem 1.2 and the explanation (4.2)) that $(x,v_i) \in M_{D_g}$ and $(x,v) \in M_{D_g}$, where x is the vector position of P. Therefore v_i and $v \in N(\dot{g}(x))$ given by (4.3). In this case (i.e., $X = R^2$) $N(\dot{g}(x))$ is a subspace of R^2 of one dimension, hence v_i and v are parallel vectors. Consequently, there is $d_i \in R$ such that $v_i = d_i v$. Combining (4.22) and (4.25) below:

$$\ddot{g}(x)(v_i)(v_i) + \dot{g}(x)(f_i(x)) = 0, \tag{4.25}$$

one obtains at once $d_i^2 = w_i(x)/w(x)$ (i.e., (4.23)), which implies (4.24) (arguing as for (4.18)).

4.3 Smooth fields of force under which a given orbit can be described

In this subsection we are concerned with the solution of the following problem (call it (P)).

(P) *Given a curve* $D_g = \{x \in A \subset R^2, \ g(x) = 0\}$ *with* g *completely smooth, find all (*g-*smooth) force field* $f:A \subset R^2 \to R^2$ *with the property that, for each* $x \in D_g$, *there is* $v = v(x,f) \in R^2$, *such that a mass particle projection from* x *with the spped* v *describes (under the action of* f*) an orbit which lies in* D_g.

In our framework, the solution of the problem)P) consists of the determination of all elements f of K_g given by (4.15).

Remark 4.4 Given a function $f:A \to R^2$, denote by f/D_g the restriction of f to D_g. Let $f \in K_g$. If $f_1:A \subset X \to X$ is g-smooth and $f_1/D_g = g/D_g$, then also $f_1 \in K_g$. This fact follows from Theorem 1.7 (or Theorem 1.8). Therefore, we are interested to determine merely the restrictions (of elements of K_g) to D_g. When there is no danger of confusion, we denote f/D_g by f (for simplicity of writing).

With the notation of §1 and

$$B = \begin{pmatrix} -g_{22} & g_{21} \\ g_{11} & -g_{11} \end{pmatrix} \tag{4.26}$$

we give the following solution to (P).

Theorem 4.3 *Let* $g:A \subset R^2 \to R$ *be completely smooth.*

(1) *If* $f \in K_g$, *then its restriction (also denoted by* f*) to* K_g *satisfies the system*

$$\begin{cases} (g_{11}g_2-g_{21}g_1)f^1(x) + (g_{21}g_2-g_{22}g_1)f^2(x) = \frac{1}{2}(w_2g_1-w_1g_2) + \frac{1}{2}\frac{a_3w(x)}{a_2} \\ g_1f^1 + g_2f^2 = w(x) \end{cases} \tag{4.27}$$

for each $x \in D_g$ (see §1).

(2) *The solution of* (4.27) *can be written in the following three equivalent forms:*

$$f(x) = -\frac{w(x)}{a_2}B\dot{g}(x) + \frac{1}{2}\frac{a_3w(x)}{a_2^2}\dot{g}(x) - \frac{1}{2a_2}\langle\dot{w}(x), \dot{g}(x)^{\perp}\rangle\dot{g}(x)^{\perp} \tag{4.28}$$

$$f(x) = -\frac{w(x)}{a_2}B\dot{g}(x) + \frac{1}{2}\frac{a_3w(x)}{a_2^2}\dot{g}(x)^{\perp} - \frac{a_1}{2a_2}\dot{w}(x) + \frac{\langle\dot{w}(x),\dot{g}(x)\rangle}{2a_2}\dot{g}(x) \tag{4.28}$$

$$f(x) = z(x)B\,\dot{g}(x) + \frac{1}{2}\|\dot{g}(x)\|^2\dot{z}(x) - \frac{1}{2}\langle\dot{z}(x),\dot{g}(x)\rangle\dot{g}(x) \tag{4.28)"}$$

for each $x \in D_g$ *where* $z:A \to R$ *is a differential function and* $z(x) > 0$ *for all*

$x \in D_G$.

(3) *Conversely, if* $w:A \to R$ *and* $z:A \to R$ *are differentiable functions such that*

$$a_2 . w(x) < 0,\ z(x) > 0,\quad \forall x \in D_g \tag{4.29}$$

and f *given by* (4.28), (4.28)' *or* (4.28)") *is locally Lipschitz, then this* $f \in K_g$.

(4) *If, in addition to the above hypotheses, we suppose that* $w,z \in C^2(A,R)$ *and* $g \in C^4(A,R)$, *then* f *given by* (4.28) - (4.28)" *belongs to* K_g *(therefore* K_g *is nonempty in this case).*

Proof (1) The first part of the theorem follows from Theorem 1.8.

(2) If, in System (4.27), w is regarded as a parameter function, then solving this elementary system we get (4.28). (First of all, one observes that the determinant of the system is just a_2 and that (1.30) holds.)

Since $y = \dot{g}(x)^{\perp} \in N_x$ (given (4.3)) and $a_2 = \ddot{g}(x)(\dot{g}(x)^{\perp})(\dot{g}(x)^{\perp})$, then (4.5) implies the first inequality of (4.29).

To prove that the form (4.28) of f is equivalent to that given by (4.28)', let us observe that, for each $x \in D_g$, there exist $a(x), b(x) \in R$ such that

$$\dot{w}(x) = a(x)\ \ \dot{g}(x) + b(x)\ \dot{g}(x)^{\perp}.$$

Namely,

$$a(x) = \frac{1}{a_1} \langle \dot{w}(x), \dot{g}(x)\rangle,\ b(x) = \frac{1}{a_1} \langle \dot{w}(x), \dot{g}(x)^{\perp}\rangle$$

where (1.18) has been used, as well as $\langle \dot{g}(x), \dot{g}(x)^{\perp}\rangle = 0$. Consequently,

$$\langle \dot{w}(x), \dot{g}(x)^{\perp}\rangle\ \dot{g}(x)^{\perp} = a_1 \dot{w}(x) - \langle \dot{w}(x), \dot{g}(x)\rangle\ \dot{g}(x), \tag{4.30}$$

which shows the equivalence of (4.28) and (4.28)'. We now prove the equivalence of (4.28)" and (4.28). Set

$$z(x) = -\frac{w(x)}{a_2},\ x \in D_g. \tag{4.31}$$

Since we have already proved the first inequality of (4.29), it follows that $z(x) > 0$, $x \in D_g$.

Let us prove that the derivative of $a_2 = a_2(x)$ in the direction $\dot{g}(x)^{\perp}$ is just a_3, i.e.,

$$\langle \dot{a}_2(x), \dot{g}(x)^{\perp} \rangle = \dddot{g}(x)(\dot{g}(x)^{\perp})(\dot{g}(x)^{\perp})(\dot{g}(x)^{\perp}) = a_3. \tag{4.32}$$

Indeed, since $a_2(x) = \ddot{g}(x)(\dot{g}(x)^{\perp})(\dot{g}(x)^{\perp})$ and

$$\langle \ddot{g}(x)(\dot{g}(x)^{\perp}), (\dot{g}(x)^{\perp})^{\bullet}(\dot{g}(x)^{\perp} \rangle = 0 \tag{4.33}$$

(which we shall prove below: $(\dot{g}(x)^{\perp})^{\bullet}$ is the derivative of $\dot{g}(x)^{\perp}$), we have

$$\langle \dot{a}_2(x), \dot{g}(x)^{\perp} \rangle = a_3 + 2\langle \ddot{g}(x)(\dot{g}(x)^{\perp}), (\dot{g}(x)^{\perp})^{\bullet} (\dot{g}(x)^{\perp}) = a_3.$$

It remains to prove (4.33). Indeed, since

$$\dot{g}(x) = \begin{pmatrix} g_1 \\ g_2 \end{pmatrix}, \dot{g}(x) = \begin{pmatrix} g_2 \\ -g_1 \end{pmatrix},$$

it follows that

$$\ddot{g}(x)(\dot{g}(x)^{\perp}) = \begin{pmatrix} \dot{g}_1(x)(\dot{g}(x)^{\perp}) \\ \dot{g}_2(x)(\dot{g}(x)^{\perp}) \end{pmatrix}, (\dot{g}(x) \ )(\dot{g}(x) \ ) = \begin{pmatrix} \dot{g}_2(x)(\dot{g}(x)^{\perp}) \\ -\dot{g}_1(x)(\dot{g}(x) \ )^{\perp} \end{pmatrix}$$

which lead us to (4.33). Obviously, (4.31) yields

$$\langle \dot{w}(x), \dot{g}(x)^{\perp} \rangle = -z(x)\langle \dot{a}_2(x), \dot{g}(x)^{\perp} \rangle - a_2(x)\langle \dot{z}(x), \dot{g}(x)^{\perp} \rangle. \tag{4.34}$$

Using (4.30) (with $\dot{z}(x)$ instead of $\dot{w}(x)$) and (4.31), (4.34) gives

$$\langle \dot{w}(x), \dot{g}(x)^{\perp} \rangle \dot{g}(x)^{\perp} = -a_3 z(x)\dot{g}(x)^{\perp} - a_2(x)a_1(x)\dot{z}(x)$$

$$+ a_2(x) \ \langle \dot{z}(x), \dot{g}(x) \rangle \dot{g}(x). \tag{4.35}$$

Combining (4.35) and (4.31), it is easy to check that (4.28) is equivalent to (4.28)".

(3) Let w and z be differentiable (on A) real-valued functions, satisfying (4.29) and f given by (4.28). Obviously,

$$\langle B\dot{g}(x), \dot{g}(x) \rangle = -a_2(x). \tag{4.36}$$

Thus, if f is given by (4.28) (or (4.28)") it follows that

$$\langle\dot{g}(x),f(x)\rangle = w(x), \quad \langle\dot{g}(x), f(x)\rangle = -a_2(x)z(x) \tag{4.37}$$

respectively.

Since $X = R^2$, any $y \in N_x$ is parallel to $\dot{g}(x)^{\perp}$, hence (4.37) and (4.29) imply (4.5). Thus, if f given by (4.28) is also locally Lipschitz, then it is a g-smooth force field.

The fact that the (unique) solution of (4.27) is given by (4.28) means that, for every w as above, f given by (4.28) satisfies (4.27) and therefore (according to Remark 4.4) we may conclude that this $f \in K_g$.

(4) In these hypotheses, f given by (4.28) is of class $C^1(A,R)$ and consequently it is locally Lipschitz. On the basis of (3) it follows that $f \in K_g$.

Remark 4.5 In the case of a general (real) Banach space X, given $g:A \subset X \to R$ completely smooth, then, in view of Theorem 4.1, the elements f of K_g are given by the g-smooth solutions of the system

$$\begin{aligned} &\ddot{g}(x)(f(x)) - \frac{a(x)}{2}\dot{g}(x) = \frac{w(x)}{2}F(x) - \frac{\dot{w}(x)}{2} \\ &\dot{g}(x)(f(x)) = w(x), \quad \forall x \in D_g. \end{aligned} \tag{4.38}$$

If in this system the unknowns are the force field f and the real valued function $a = a(x)$, while w is regarded as a parameter function, then by Proposition 4.3 it follows that (4.38) has a unique solution (at least in the case $X = R^n$). Indeed, if $X = R^n$ the matrix of the linear system (4.38) is given by the nonsingular matrix (4.14)". The formula in (4.28) can also be established via the linear system (4.38).

4.4 Orbital uniform motions in Hilbert spaces

Let H be a real Hilbert space of inner product $\langle\cdot,\cdot\rangle$ and norm $\|\cdot\|$, and let S(r) be the sphere of radius r about 0 (see (2.37)). In this case the first result we need is:

Corollary 4.2 *The subset given by* (1.12) *becomes*

$$M_{S(r)} = \{(x,y) \in A \times H;\ \|x\| = r,\ \langle x,y\rangle = 0,\ \|y\|^2 + \langle x,f(x)\rangle = 0\}. \tag{4.39}$$

Proof One applies Proposition 2.1 with $z = f(x)$. For a direct proof, see Remark 4.6.

Corollary 4.3 *Suppose that* $A \supset S(r)$ *is an open subset of* H. *Let* $f:A \to H$ *be a locally Lipschitz function such that:*

(1) $\langle x,f(x)\rangle < 0, \quad x \in S(r)$.

(2) *The real valued function* $w(x) = \langle x,f(x)\rangle$ *is differentiable on* A. *Then* $S(r)$ *is flow-invariant for* (1.3) *iff*

$$2\langle f(x),y\rangle + \dot{w}(x)(y) = 0, \quad \forall (x,y) \in M_{S(r)}. \tag{4.40}$$

Proof One uses Theorem 1.7 with $g(x) = \frac{1}{2}(\|x\|^2 - r^2)$. In this case the invariance condition (1.13) becomes (4.40).

We now proceed to the definition and characterization of the uniform motions on $S(r)$.

Definition 4.3 *The motion on* $S(r)$ *under the force field* f *is said to be uniform, if there is a constant* $k = k(f,r) > 0$ *such that, for every* $x \in S(r)$ *and* $y \in H$ *with* $\|y\| = k$, *and* $\langle x,y\rangle = 0$, *the unique solution to* (1.3) *satisfies*

$$\|u(t)\| = r, \quad \forall t = 0. \tag{4.41}$$

The next theorem shows that (as in the classical case of R^3) the only force field on $S(r)$ under which the motion on $S(r)$ is uniform is the central force field.

Theorem 4.4 *Let* $f:S(r) \subset H \to H$. *The motion on* $S(r)$ *is uniform if and only if*

$$f(x) = -\frac{k^2}{r^2}x, \quad \forall x \in S(r). \tag{4.42}$$

Proof *The necessity of* (4.42). Suppose that the motion on $S(r)$ under f is uniform in the sense of Definition 4.3. Then, for each $x \in S(r)$, there is $y \in H$ such that

$$\langle x,f(x)\rangle = \|y\|^2 = -k^2, \quad \langle x,y\rangle = 0. \tag{4.43}$$

Indeed, following the proof of Theorems 1.2 and 1.3 we conclude that $(u(t), u'(t)) \in M_{S(r)}$ with

$$M_{S(r)} = \{(x,y);\ \lim_{h\downarrow 0} h^{-2}\, d(x + hy + \frac{h^2}{2} f(x);\ S(r)) = 0\}. \tag{4.44}$$

Alternatively, it can be proved directly (without Formula (1.12)) that (4.44) is equivalent to (4.39). In other words, if the motion on S(r) is uniform, then $\|u'(t)\| = k,\ \ \forall t \geq 0$,

$$M_{S(r)} = \{(x,y) \in H \times H;\ \|x\| = r,\ \|y\| = k,\ \langle x,y\rangle = 0\} \tag{4.45}$$

and (1.6) holds (with S(r) in place of D). In view of Lemma 2.3, (1.6) is equivalent to

$$\lim_{h\downarrow 0} d(y + hf(x);\ S(k)) = 0,\quad \forall (x,y) \in M_{S(r)}. \tag{4.46}$$

Finally, by Proposition 2.1, the tangential condition (4.46) is equivalent to

$$\langle f(x),y\rangle = 0,\quad \forall x \in S(r),\quad y \in H,\ \langle x,y\rangle = 0. \tag{4.46)'$$

Now let us fix $x \in S(r)$. Set $H_x = \{ax;\ a \in R\}$. If we denote by $H_x^{\perp}$ the orthogonal subspace of H_x then $H = H_x \oplus H_x^{\perp}$ (the direct sum). It follows that $f(x) \in H_x$, that is, $f(x) = ax$, with $a \in R$. Since $\langle f(x),x\rangle = -k^2$, and $\langle x,x\rangle = r^2$, one obtains (4.42).

The sufficiency of (4.42). If f is given by (4.42) then $M_{S(r)}$ is nonempty and (4.46) is satisfied. Then, on the basis of Theorem 1.4 with S(r) in place of D and $M_{S(r)}$ given by (4.44), we get (4.41). The proof is complete.

Remark 4.6 In Theorem 4.4, f is defined on S(r) only. Without Formula (1.12), the equivalence of $M_{S(r)}$ given by (4.44) with (4.39) can be proved as follows. Suppose that (4.44) holds. Then by Lemma 2.2 there is $u(h) \to 0$ as $h \downarrow 0$ such that

$$b(h) = x + hy + \frac{h^2}{2}(f(x) + u(h)) \in S(r),\quad \forall h > 0. \tag{4.46)''$$

Now $\|b(h)\|^2 = r^2$ implies (4.39).

Conversely, if (4.39) holds, then we can find u(h) satisfying (4.46)". Indeed, choose u(h) as below:

$$x + hy + \frac{h^2}{2}(f(x) + u(h)) = r(x + hy + \frac{h^2}{2} f(x))/(\|x + hy + \frac{h^2}{2} f(x)\|).$$

Then (4.49) yields $u(h) \to 0$ as $h \downarrow 0$ and thus we get (4.44).

In view of Theorem 1.3 and of (4.45), Definition 4.3 can be restated as:

Definition 4.3' *The motion on* $S(r)$ *under the force field* f *is uniform, if there is a constant* $k = k(f,r) > 0$ *such that* $S(r)$ *is flow-invariant with respect to* (1.3) *and* $\|u(t)'\| = k$, $\forall t \geq 0$.

The necessity of (4.46)' can also be derived as follows. On the basis of (1.3) we have

$$\frac{1}{2}\frac{d}{dt}\|u'(t)\|^2 = \langle f(u(t)), u'(t)\rangle, \quad \forall t \geq 0.$$

It is now clear that this formula and the hypothesis that the motion on $S(r)$ is uniform, imply (4.46)'.

Let us apply Theorem 4.4 to the problem of the launching of an Earth satellite in a circular orbit (one assumes that the oblateness of the Earth, air resistance, the attraction of other celestial bodies etc. are neglected). We apply the previous results to the Newtonian gravitational field:

$$f(x) = - GM\, x/\|x\|^2. \tag{4.47}$$

Here GM is a positive constant (it is called the power of the force centre). In this case $H = R^3$, $S(r)$ is the sphere of radius r about Earth's centre 0 (as the force centre) and $r = R + r_0$, where R is the radius of the Earth, while $r_0 > 0$.

As a consequence of Theorem 4.4 one obtains the following classical result of flight space:

Corollary 4.4 *If a body is projected from point* x_0 *at a distance* $r = R + r_0$ *from the Earth's centre, with speed* y_0 *(parallel to the Earth's surface, i.e.,* $\langle x_0, y_0\rangle = 0$) *of magnitude*

$$\|y_0\| = (GM/(R + r_0))^{1/2} \tag{4.47'}$$

then the body describes (uniformly) the circle

$$C(r) = P(x_0, y_0) \cap S(R + r_0),$$

where $P(x_0,y_0)$ *is the plane spanned by the vectors* x_0 *and* y_0.

Proof In this case we have $\langle x,f(x)\rangle = -GM/r$, $\forall x$ with $\|x\| = r = R + r_0$, therefore Theorem 4.4 holds with $k^2 = GM/r$. Obviously, $(x_0,y_0) \in M_{C(r)}$ given by (4.45) with $r = R + r_0$. It follows that C(r) is a flow-invariant set, which concludes the proof.

Other applications of this type will be given in the sequel.

4.5 Flow invariance of the conic in a Newtonian field

We apply Theorem 1.8 in the case of the conic

$$C = \{x \in R^2;\ \|x-\bar{a}\|^2 - (\langle\bar{b},x-\bar{a}\rangle + d)^2 = 0\}, \tag{4.48}$$

where $\bar{a},\bar{b} \in R^2$, $d \in R$ and $d \neq 0$, $\bar{a}$ being one of the foci.

In this case $C = D_g$ with g given by (4.9). A general result on C is given by Proposition 4.1. First we discuss (1.24) for each particular conic and f given by (4.47), and then we verify (1.13). We have already considered the circle C(r) in Corollary 4.4.

4.5.1 Elliptic velocity

Let us consider the ellipse

$$E = \left\{x = \binom{x_1}{x_2} \in R^2,\ \frac{(x_1+c)^2}{a^2} + \frac{x_2^2}{b^2} - 1 = 0\right\} \tag{4.49}$$

where the rectangular axes are taken through one of the foci, 2a (resp. 2b) is the major (minor) axis and $a^2 - b^2 = c^2$. Denote by O the intersection of axes and

$$g(x) = \frac{1}{2}\left(\frac{(x_1+c)^2}{a^2} + \frac{x_2^2}{b^2} - 1\right), \text{ (hence } D_g = E). \tag{4.50}$$

Lemma 4.1 (1) *Let* $A = R^2 - \{0\}$. *With* f *given by* (4.47), (1.24) *and* (1.11) *become respectively*

$$M_E = \left\{(x,y) \in E \times R^2,\ \frac{(x_1+c)y_1}{a^2} + \frac{x_2y_2}{b^2} = 0,\ y_1^2 + y_2^2 = GM\left(\frac{2}{r} - \frac{1}{a}\right)\right\} \tag{4.51}$$

where y_1,y_2 *are the coordinates of* y *and* $r = (x_1^2 + x_2^2)^{1/2} = \|x\| < 2a$;

$$w(x) = \dot{g}(x)f(x)) = -\frac{GM}{ar^2}, \; x \in E. \qquad (4.52)$$

Proof It is easy to check that

$$r = \frac{1}{a}(b^2 - cx_1), \; a(r-a) = -c(x_1 + c), \; x \in E. \qquad (4.53)$$

In this case (with the notation of (1.15) - (1.22),

$$g_1(x) = \frac{x_1+c}{a^2}, \; g_2(x) = \frac{x_2}{b^2}, \; g_{11} = \frac{1}{a^2}, \; g_{22} = \frac{1}{b^2}, \; g_{12} = 0 \qquad (4.54)$$

and therefore

$$a_1(x) = \frac{1}{a^2b^2}(2ar-r^2), \; a_2(x) = \frac{1}{a^2b^2}, \; x \in E. \qquad (4.55)$$

Finally,

$$\dot{g}(x)(f(x)) = -\frac{GM}{r^3}\left(\frac{x_1(x_1+c)}{a^2} + \frac{x_2^2}{b^2}\right) = -\frac{GM}{ar^2}, \; x \in E. \qquad (4.56)$$

Combining (4.54), (4.55), (4.56), (1.22) and (1.24), one obtains (4.51).

4.5.2 Hyperbolic velocity

If

$$g(x) = \frac{1}{2}\left(-\frac{(x_1+c)^2}{a^2} + \frac{x_2^2}{b^2} + 1\right), \; x = \begin{pmatrix} x_1 \\ x_2 \end{pmatrix}, \qquad (4.57)$$

then D_g is the hyperbola

$$H = \{x \in R^2 - \{0\}, \; \frac{(x_1+c)^2}{a^2} - \frac{x^2}{b^2} - 1 = 0\}, \qquad (4.58)$$

where the rectangular axes are taken through one of the foci 0(0,0) and

$$r = \|x\| = \frac{1}{a}|b^2 + cx_1|, \quad x \in H, \quad c^2 = a^2 + b^2. \qquad (4.59)$$

It is necessary to consider the following two cases (for A):

$$A_1 = \{x \in R^2 - \{0\}, x_1 > -\frac{b^2}{c}\} \qquad (4.60)$$

$$A_2 = \{x = \begin{pmatrix} x_1 \\ x_2 \end{pmatrix} \in R^2, \; x_1 < -\frac{b^2}{c}\}. \qquad (4.61)$$

Denote by $H_i = H \cap A_i$ the branch of H which is contained in A_i, $i = 1,2$. Obviously 0 is regarded as the centre of the force field (4.47) and H_1 is the branch of H around this centre. We shall prove (in (4.5.4)) that H_1 is a flow-invariant set under f given by (4.47), while H_2 has this property under a "repulsive" field (of centre 0),

$$\bar{f}(x) = GM \frac{x}{\|x\|^3}, \; x \in A_2. \tag{4.62}$$

This type of field (4.62) (of the inverse square repulsion $K/\|x\|^2$, $K > 0$ per unit mass) is useful in physics for the bombardment of atoms by α-particles.

Lemma 4.2 (1) *With* f *given by* (4.47) *and* $D_g = H_1$, (1.24) *and* (1.11) *become, respectively,*

$$M_{H_1} = \{(x,y) \in H_1 \times R^2, \frac{y_1(x_1+c)}{a^2} - \frac{y_2 x_2}{b^2} = 0, \; y_1^2 + y_2^2 = GM(\frac{2}{r} + \frac{1}{a})\} \tag{4.63}$$

and

$$\dot{g}(x)(f(x)) = -\frac{GM}{a^2 r^3}(b^2 + cx_1) = -\frac{GM}{ar^2}, \; x \in H_1 = H \cap A_1. \tag{4.64}$$

(2) *In the repulsive case* (4.62) *and* $D_g = H_2$,

$$M_{H_2} = \{(x,y) \in H_2 \times R^2, \frac{y_1(x_1+c)}{a^2} - \frac{y_2 x_2}{b^2} = 0, \; y_1^2 + y_2^2 = GM(\frac{1}{a} - \frac{2}{r})\} \tag{4.65}$$

and

$$\dot{g}(x)(\bar{f}(x)) = \frac{GM}{a^2 r^3}(b^2 + cx_1) = -\frac{GM}{ar^2}, \; x \in H_2 = H \cap A_2. \tag{4.66}$$

Proof (1) First of all, in this case

$$r = \frac{1}{a}(b^2 + cx_1), \; a(a + r) = c(c + x_1), \; x \in H_1$$

$$g_1(x) = -\frac{x_1 + c}{a^2}, \quad g_2(x) = \frac{x_2}{b^2}, \tag{4.67}$$

and (4.64) follows at once. Further, by (4.67) we have (for $x \in H_1$)

$$a_1(x) = g_1^2 + g_2^2 = \frac{(x_1+c)^2}{a^4} + \frac{x_2^2}{b^4} = \frac{1}{a^4b^2}(c^2(x_1+c)^2 - a^4)$$

$$= \frac{1}{a^2b^2}(r^2 + 2ar) \tag{4.68}$$

$$a_2(x) = \frac{1}{a^2b^2} \tag{4.68)'}$$

which yield (4.63).

(2) In this case $b^2 + cx_1 < 0$, therefore (4.59) gives

$$r = -\frac{1}{a}(b^2 + cx_1) \text{ (hence } a(a-r) = -c(c + x_1)), \; x \in H_2. \tag{4.69}$$

Therefore

$$a_1(x) = \frac{1}{a^4b^2}[c^2(x_1+c)^2 - a^4] = \frac{1}{a^2b^2}(r^2 - 2ar), \; x \in H_2$$

while a_2 is given by (4.68)'.

According to (4.69), we easily get (4.66). Replacing a_1, a_2 and $\dot{g}(x)(\bar{f}(x)$ in (1.24), (4.65) follows.

4.5.3 <u>Parabolic velocity</u> (of escape from the force centre)

Denote by P the parabola

$$P = \{x \in R^2, \; x_2^2 = p^2 + 2p\, x_1\} \tag{4.70}$$

where p is the distance from the focus 0 (as the origin of the rectangular axes) to the directrix. Clearly, $P = D_g$ with $g(x) = \frac{1}{2}(x_2^2 - 2px_1 - p^2)$. Using the elementary fact that in this case

$$r = \|x\| = p + x_1, \text{ if } x = \begin{pmatrix} x_1 \\ x_2 \end{pmatrix} \in P,$$

we get immediately (for f given by (4.47))

$$\dot{g}(x)(f(x)) = -GM\, p/r^2, \; a_1 = 2pr, \; a_2 = p^2. \tag{4.71}$$

In view of (4.71), (1.24) becomes

$$M_P = \{(x,y) \in P \times R^2; \; py_1 = x_2y_2, \; y_1^2 + y_2^2 = \frac{2CM}{r}\}. \tag{4.72}$$

The speed y of a mass particle (at the point $x \in P$) such that $(x,y) \in M_p$ (therefore having the magnitude $y^2 = 2GM/r$, $r = \|x\|$) is called "the velocity of escape from the force centre" (cf. McCuskey [1], p. 27).

4.5.4 Flow invariance of the conic under a Newtonian field

The general equation of the conic C is given by (4.48). We are now able to prove (via the flow-invarinace method) the following result.

Theorem 4.5 (1) *Any conic* C *is a flow-invariant set for Equation* (1.3) *with*

$$f(x) = -GM\frac{x-\bar{a}}{\|x-\bar{a}\|^3}, \quad x \neq \bar{a} \tag{4.73}$$

where, in the case of hyperbola, $C = H_1$ *(i.e., the branch around the focus* $\bar{a}$ *as the centre of the attractive field* (4.73)).

(2) *The other branch* H_2 *of the hyperbola is flow-invariant for* (1.3) *with the repulsive field*

$$f(x) = GM\frac{x-\bar{a}}{\|x-\bar{a}\|^3}, \quad x \neq \bar{a} \tag{4.74}$$

where GM *is the power of the force centre* $\bar{a}$.

Proof We apply Theorem 1.8. We have already seen that in this case M_D (with g given by(4.9)) is nonempty (actually, that (1.23) holds). This fact is proved by (4.40), (4.44), (4.51), (4.63), (4.65) and (4.72). It remains to verify (1.25) or equivalently (1.13). In this case $\dddot{g}(x) = 0$ and f is differentiable on $R^2 - \{\bar{a}\}$, therefore (1.13) becomes

$$3\ddot{g}(x)(f(x))(y) + \dot{g}(x)(\dot{f}(x)(y)) = 0, \tag{4.75}$$

for all (x,y) with $g(x) = 0$, $\dot{g}(x)(y) = 0$. Indeed, in this case the derivative of w (given by (1.11)) in the direction y is the following:

$$\dot{w}(x)(y) = \ddot{g}(x)(f(x))(y) + \dot{g}(x)(\dot{f}(x)(y)),$$

and $N(\dot{g}(x))$ defined by (4.3) is a subspace of one dimension. Let us observe that the derivative of the function $h(x) = \|x-\bar{a}\|$, $x \in R^2$, $x \neq a$ is just

$$\dot{h}(x)(y) = \frac{\langle x-\bar{a},y\rangle}{\|x-\bar{a}\|}.$$

Using this remark as well as (4.11), (4.12) and (4.13), we have successively

$$\dot{f}(x)(y) = \frac{GM\ y}{||x-\bar{a}||^3} + \frac{3GM\langle x-\bar{a},y\rangle}{||x-\bar{a}||^5}(x-\bar{a}),$$

$$I = 3\ddot{g}(x)(f(x))(y) + \dot{g}(x)(\dot{f}(x)(y) = 3\langle f(x),y\rangle - 3\langle\bar{b},f(x)\rangle\ \ \langle\bar{b},y\rangle$$

$$- GM\ \dot{g}(x)(y)\ \frac{1}{||x-\bar{a}||^3} + \frac{3GM\langle x-\bar{a},y\rangle}{||x-a||^2}\ \dot{g}(x)(x-\bar{a})$$

$$= -\frac{3GM}{||x-\bar{a}||^3}(\langle x-\bar{a},y\rangle - \langle\bar{b},x-\bar{a}\rangle\langle\bar{b},y\rangle)$$

$$+ \frac{3GM\ d\langle x-\bar{a},y\rangle}{||x-\bar{a}||^5}(d + \langle\bar{b},x-\bar{a}\rangle),$$

where we have also used

$$\dot{g}(x)(x-\bar{a}) = d(d + \langle\bar{b},x-\bar{a}\rangle),\quad \forall x \in C.$$

Since $\dot{g}(x)(y) = 0$ gives

$$\langle x-\bar{a},y\rangle - \langle\bar{b},x-\bar{a}\rangle\ \langle\bar{b},y\rangle = d\langle\bar{b},y\rangle,\quad \forall x \in C,$$

I becomes,

$$I = -\frac{3GM\ d}{||x-\bar{a}||^5}\ [\ ||x-\bar{a}||^2\ \langle\bar{b},y\rangle - \langle x-\bar{a},y\rangle\ (d + \langle\bar{b},x-\bar{a}\rangle)$$

$$= \frac{3GM\ d}{||x-\bar{a}||^5}(d + \langle\bar{b},x-\bar{a}\rangle)\dot{g}(x)(y) = 0$$

for $g(x) = 0$ and $\dot{g}(x)(y) = 0$. Hence (4.75) holds for f given by (4.73). Since $\bar{f} = -f$, the proof is complete.

4.5.5 <u>Motion in a central field of force</u>

Recall that $f:A \subset R^3 \to R^3$ (see 4.1) is said to be a "central force field" of centre 0 if, for each vector position $x \in A$, the force f(x) (associated with x) acts along the vector which joins 0 with x. The central force field is said to be "attractive" ("repulsive") if, for each $x \in A$, the vector force f(x) is directed toward (resp. away) from 0.

In terms of dynamics, Theorem 4.5 asserts that a body P projected from a point $x \in C$ with a velocity y such that $(x,y) \in M_C$ (given by (4.51) etc.)

describes (under the action of the field (4.73) the orbit C. However, in the context of dynamics much more is known, namely the following famous result holds:

<u>Theorem 4.6</u> (1) *A mass particle* Q *moving under a central field of force describes an orbit which lies in a plane.*

(2) *If* Q *describes the conic* C *with constant real velocity, then the force acting on it varies inversely as the square of the distance from* Q *to the focus* $\bar{a}$ *of* C.

(3) *Conversely, if* Q *is projected from any point* $x \in C$ *with speed* y *such that* $(x,y) \in M_C$, *given by* (4.44) (*with* $k = (\frac{GM}{r})^{1/2}$, $r = \|x\|$), (4.51), (4.63), (4.65), (4.72), *describes* (*under the action of given by* (4.73) *or* (4.74)), *the conic* C *with constant areal velocity* (*relative to the focus* $\bar{a}$.).

In our framework (i.e., via the flow-invariance method) this theorem can easily be proved as follows:

(1) Let Q move under the central force field f of centre O. Denote by x_0 (resp. y_0) the initial position (velocity) of Q and by S the closed linear subspace of R^3 spanned by the vectors x_o and y_o. If $y_o \neq 0$ and y_o is not parallel to x_0 then S is a plane (otherwise S is a straight line containing $x_0 \neq 0$). Let $A = R^3 - \{0\}$ and $x \in D = A \cap S$. Since f(x) acts along the vector which joins x and O and x, $0 \in S$, it follows that $f(x) \in S$ (i.e., $f(D) \subset S$) and therefore (by Theorem 1.5) D is a flow-invariant set for (1.3). Obviously, $D = \bar{D} \cap A$ (i.e., D is closed in A) since $D = S - \{0\}$ and hence $\bar{D} = S$. Inasmuch as $(x_0,y_0) \in D \times S = M_D$, in view of Theorem 1.3 the orbit $u = u(t)$, $t \geqslant 0$ of Q lies in D, hence in S.

(2) For the sake of simplicity we make a choice, supposing, e.g., that C is the ellipse E (4.49).

Therefore let Q describe E with constant areal velocity (cf. McCuskey [1] p. 5) relative to the focus O as the origin of the rectangular axes. First of all, this implies that the force field f acting on it is central, i.e., for each $x \in E$ there is $h(x) \in R$ such that

$$f(x) = h(x)x. \tag{4.76}$$

If $u = (u_1(t), u_2(t))$ is the law of motion of Q, then the constancy of areal velocity (relative to the focus O) means (cf. McCuskey [1], p. 6) that

$$u_1(t)\, u_2'(t) - u_2(t)\, u_1'(t) = K = \text{const},\ t \geqslant 0. \tag{4.77}$$

Since u is E-valued then $(u(t), u'(t)) \in M_E$ (Theorem 1.2), therefore

$$\begin{cases} \dfrac{y_1(x_1+c)}{a^2} + \dfrac{x_2 y_2}{b^2} = 0 \\ x_1 y_2 - x_2 y_1 = K \end{cases} \tag{4.78}$$

for all $(x,y) \in M_E$ given by (4.51), where we have denoted $x_i = u_i(t)$ and $y_i = u_i'(t)$, $i = 1,2$.

Using some of (4.47) - (4.53), one observes that the solution (y_1, y_2) of (4.78) is the following:

$$y_1 = -\frac{aKx_2}{rb^2},\quad y_2 = \frac{K(x_1+c)}{ar},\ (r = \|x\|,\ x \in E)$$

and hence

$$y^2 = y_1^2 + y_2^2 = \frac{K^2}{b^2 r^2}(2ar - r^2),\quad x \in E. \tag{4.79}$$

On the other hand, by (1.24), (4.74) and (4.76) we have

$$\begin{aligned} y^2 &= -\frac{a_1}{a_2}\langle \dot{g}(x), f(x)\rangle = -h(x)\frac{a_1}{a_2}(g_1 x_1 + g_2 x_2) \\ &= -h(x) r(2ar - r^2)/a \end{aligned} \tag{4.79$'$}$$

for each $x \in E$. Comparing (4.79) and (4.79)' we conclude that

$$h(x) = -\frac{aK^2}{b}\frac{1}{r^3}\ (r = \|x\|). \tag{4.80}$$

(3) We now assume that f is given by (4.47). If Q is projected from any point $x \in C$ as mentioned in the theorem, then Q describes under f, S(r), E, H_1 and P respectively (and H_2 under repulsion (4.62)). This aspect has been already proved (Theorem 4.5, as well as Corollary 4.4 in the case C = S(r)). It remains to prove that Q obeys the law of areas (i.e., that the areal velocity relative to the focus 0 is constant). Indeed, the areal velocity (denote it by 2K(t)) is given by

$$K(t) = u_1(t)\, u_2'(t) - u_2(t)\, u_1'(t)$$

as we have seen above (see (4.77)).

In the case of E, (4.80) also holds with $h(x) = -GM/r^3$ (by (4.47)), therefore

$$K^2(t) = -r^3 bh(x)/_a = GM\, b/_a = \text{const.}$$

4.5.6 The determination of smooth force fields under which a conic can be described

Applying Theorem 4.3, we can derive all g-smooth force fields under which the conic C can be described (with g given by (4.9)). For simplicity we treat this problem in the case of C(r) only, where

$$C(r) = \{x \in R^2, \|x\| = r\}, \; r > 0. \tag{4.81}$$

The rest of the discussion is left to the reader. Obviously, C(r) corresponds to D_g with

$$g(x) = \frac{1}{2}\left(\|x\|^2 - r^2\right) = \frac{1}{2}\left(x_1^2 + x_2^2 - r^2\right).$$

In this case, $\dot{g}(x) = x$ and (4.26) becomes

$$B = \begin{pmatrix} -1 & 0 \\ 0 & -1 \end{pmatrix}.$$

In view of Theorem 4.3 (Formula (4.28)"), we get:

Corollary 4.5 (1) *All force fields under which* C(r) *can be described are given by*

$$f(x) = -z(x)x + \frac{r^2}{2}\dot{z}(x) - \frac{1}{2}\langle x, \dot{z}(x)\rangle x, \tag{4.82}$$

where $z: R^2 - \{0\} \to R$ *is a continuously differentiable function and*

$$z(x) > 0 \text{ for } \|x\| = r.$$

(2) *If in addition we assume that the function* z *is positively homogeneous of degree* 0 (*i.e.*, $z(tx) = z(x),\ \forall t > 0$), *then for each natural number* n, *the sphere* $S(r) = \{x \in R^n,\ \|x\| = r\}$ *can be described under the force field*

$$f(x) = -z(x)x + \frac{r^2}{2}\dot{z}(x),\; x \in R^n,\; x \neq 0. \tag{4.83}$$

Proof (1) This assertion follows directly from (4.28)", as we have already mentioned.

(2) In the case $n = 2$, this assertion follows from (4.82) and Euler's theorem on continuously differentiable, positively homogeneous functions (of degree zero), namely

$$\langle x, \dot{z}(x) \rangle = 0, \ x \in R^n (\dot{z}(x) = \operatorname{grad} z(x)). \tag{4.84}$$

For $n \geq 3$, we have to prove it. First of all, for f given by (4.83), the subset $M_{S(r)}$ given by (4.40) is the following:

$$M_{S(r)} = \{(x,y) \in S(r) \times R^n, \ \langle x,y \rangle = 0, \ \|y\|^2 = r^2 z(x)\} \tag{4.85}$$

where (4.84) has been used.

We have to prove that any solution u of the equation

$$u'' = -z(u)u + \frac{r^2}{2} \dot{z}(u)$$

with $(u(0), u'(0)) \in M_{S(r)}$) is S(r)-valued (i.e., $u(t) = r$ as long as it exists). To do this, set

$$v(t) = \frac{1}{2} (\| u(t) \|^2 - r^2).$$

Using (4.85) it is easy to check that v satisfies the linear (scalar) differential equation

$$v'''(t) = -4z(u(t))v'(t) - 2 \langle u'(t), \dot{z}(u(t)) \rangle v(t) \tag{4.86}$$

and the initial conditions

$$v(0) = 0, \ v'(0) = \langle x,y \rangle = 0, \ v''(0) = \|y\|^2 - r^2 z(x) = 0$$

where $u(0) = x$, $u'(0) = y$ (i.e., $(x,y) \in M_S(r)$. Therefore, v is the trivial solution of (4.86), which implies $\|u(t)\| = r$, $t \geq 0$.

Remark 4.7 (1) The notion "S(r) can be described under f" is obviously interpreted as "S(r) is a flow-invariant set for (1.3)" (see (4.1) and (4.2)).

(2) By Theorem 4.4 it follows that the motions on S(r) under force field (4.83) are uniform iff

$$\langle x,f(x)\rangle = -r^2z(x) = -k^2 = \text{const}, \quad \forall x \in S(r)$$

(i.e., iff f is the attractive field $f(x) = -kx$).

(3) According to (1.27) it follows that f given by (4.82) is the solution of the partial differential equation

$$3x_2f^1 - 3x_1f^2 + x_1x_2\left(\frac{\partial f^1}{\partial x_1} - \frac{\partial f^2}{\partial x_2}\right) - x_1^2\frac{\partial f^1}{\partial x_2} + x_2^2\frac{\partial f^2}{\partial x_1} = 0 \tag{4.87}$$

where

$$x_1^2 + x_2^2 = r^2,$$

which characterizes the continuously differentiable force field $f = \begin{pmatrix} f_1 \\ f_2 \end{pmatrix}$ under which C(r) can be described.

§5 Notes and Remarks

The main steps of this chapter are the following:

(1) Introduction of the subset M_D given by (1.4).

(2) Definition 1.2 of the invariance of D with respect to the equation $u'' = f(u)$ (in other words, with respect to the force field f acting on D).

(3) The tangential condition (1.6), which may be called the invariance condition of the subset D under the force field f.

(4) Characterization of M_D and (1.6).

(5) Application of the general results to orbital motion.

The first main remark consists of the fact that, for the flow-invariance (in short: invariance) of D under f, the initial "speed" y in (1.3) must be chosen such that $(x,y) \in M_D$ (Theorem 1.2). This condition is not sufficient (see Theorem 1.4). In some particular cases, the invariance condition (1.6) is a partial differential equation (see (1.26) - (1.27)). Integratint this partial differential equation we get all g-smooth force fields f (i.e., (4.28)) under which $D = D_g = \{x, g(x) = 0\}$ can be described. By the classical methods of mechanics, similar forms of f have been obtained by Dainelli (1880). Let us note that the formula

$$(*) \quad \|y\|^2 = -\frac{\dot{g}(x)(f(x))}{c(x)\,\|\dot{g}(x)\|} \quad (x(x) \text{ being the curvature of } D_g \text{ at } x),$$

which appears in (1.24), is the square magnitude of the speed of "projection" from $x \in D_g$. More precisely, if f satisfies the invariance condition (1.27), then a mass particle projected from $x \in D_g$ with speed y (which is tangent to D_g at x) of square norm (*), describes an orbit which lies in D_g. In particular, when D_g is a conic and $f = f_N$ is the Newtonian field (4.73), the formula (*) is a cosmic speed. See (4.77)', (4.51), (4.63), (4.64) and (4.72). By classical methods, these speeds are discussed in all textbooks of dynamics (see, e.g., McCuskey [1, p. 27], Pars [1, pp. 28,29], Whittacker [1, p. 88]). Moreover, in this particular case, the invariance condition (1.25) is fulfilled (Theorem 4.5). Consequently, in our present framework, a conic is a flow-invariant set with respect to the force field f varying inversely as the square of the distance.

Finally, our proof here of the equivalence of the first two laws of Kepler with the Newtonian field f_N, is an "algebraic" proof (see the proof of Parts (2) and (3) of Theorem 4.6).

The idea of considering M_D goes back to the author [1], and the author and Ursescu [2,3]. Actually, this chapter follows the author and Ursescu [3].

We have proved by means of tangent sets that Theorem 1.4 (on the invariance of D under f) has a strong unifying effect in flight space. This technique of tangent sets has also been used in optimization (see, e.g., §6 in Chapter 4). By using such a technique, Clarke [3] and Ursescu have obtained significant extensions of the Pontrygin maximum principle and of the Lagrange multipliers rule. The flow-invariance method has been applied by Bourguignon and Brezis (1974) to the Euler equation and by Abraham and Marsden (1978) to Hamiltonian systems.

By means of the transversality approach (in place of surjectivity), some of the ideas of this chapter are extended to the Banach manifolds in Chapter 4.

4 Flow invariance on Banach manifolds and some optimization problems

§1. Quasi-tangent vectors to a subset S of a Banach manifold

Let M be a Hausdorff C^k-manifold, $k \geq 1$, which is modelled by a Banach space E, and let S be a nonempty subset of M. Denote by $T_x(M)$ the tangent space of M at $x \in M$ and by $T(M) = \bigcup_{x \in M} T_x(M)$ the tangent bundle of M. For the notation and some basic properties concerning the manifolds we refer to Lang [1] and Palais [1]. However, for the sake of completeness, let us recall that T(M) is a C^{k-1}-manifold with respect to the following differentiable structure: let $\pi: T(M) \to M$ be the canonical projection of T(M) onto M (i.e., $\pi(T_x(M)) = x$). If $\phi: U \subset M \to E$ is a chart of M, then $\tilde{\phi}: \pi^{-1}(U) \subset T(M) \to E \times E$ given by

$$\tilde{\phi}(y) = (\phi(\pi(y)), D(\phi)_{\pi(y)} \cdot y),\ y \in \pi^{-1}(U) \tag{1.1}$$

is a chart of T(M). As usual, the norm of E is denoted by $\|\cdot\|$.

The main problem now consists of the extension of the tangential condition (2.1) in Chapter 3 from a Banach space to a Banach manifold. In this direction we give:

Definition 1.1 *A vector* $y \in T_x(M)$ *is said to be quasi-tangent to the subset* $S \subset M$ *at* $x \in S$ *(to the right) if there is a chart* (U,ϕ) *of* M *at* x *such that*

$$\lim_{h \downarrow 0} \frac{1}{h} d(\phi(x) + hD(\phi)_x y;\ \phi(U \cap S)) = 0 \tag{1.2}$$

(*where* d(u,A) *stands for the distance in* E *from* $u \in E$ *to the subset* $A \subset E$).

Of course we have to prove that the notion of quasi-tangency is independent of the local chart (U,ϕ).

Proposition 1.1 *If there is a chart* (U,ϕ) *such that* (1.2) *holds, then for every chart* (V,ψ) *at* x *we have a similar property, i.e.*

$$\lim_{h \downarrow 0} \frac{1}{h} d(\psi(x) + hD(\psi)_x y;\ \psi(V \cap S)) = 0. \tag{1.3}$$

Proof In view of the proof of (3.9)' in Chapter 2, it follows that (1.2) is equivalent to the existence of a function $r_\phi \equiv r:R_+ \to E$ with $r(h) \to 0$ as $h \downarrow 0$ such that

$$a_h = \phi(x) + hD(\phi)_x y + hr(h) \in \phi(U \cap S), \quad \forall h \in R_+. \tag{1.4}$$

Consequently, $\psi\phi^{-1}(a_h) \in \psi(U \cap V \cap S)$ for all sufficiently small $h > 0$. By the definition of Fréchet derivative (denoted by D) and according to (1.4), there is $\tilde{r}(h) \to 0$ as $h \downarrow 0$ with the property

$$\psi\phi^{-1}(a_h) = \psi(x) + hD(\psi\phi^{-1})_{\phi(x)}(D(\phi)_x y + r(h)) + h\tilde{r}(h). \tag{1.5}$$

On the other hand, $D(\psi)_x y = D(\psi\phi^{-1})_{\phi(x)}(D(\phi)_x y)$ and therefore (1.5) implies

$$\lim_{h\downarrow 0} h^{-1} \|\psi(x) + hD(\psi)_x y - \psi\phi^{-1}(a_h)\| = 0$$

which yields (1.3). The proof is complete.

In other words, we now have a notion of tangency to a subset S of M, which is not a submanifold of M. Moreover, in the case in which S is a submanifold of M, then quasi-tangency means just tangency in the classical sense. More precisely, we have:

Theorem 1.1 *Let* S *be a* C^k*-submanifold of* M, $k \geq 1$, *and let* $x \in S$ *and* $y \in T_x(M)$. *Then* y *is quasi-tangent to* S *at* x *iff* $y \in T_x(S)$.

Proof Since S is a submanifold of M, there is a closed linear subspace E_1 of E such that $E = E_1 \oplus E_2$ (topological direct sum) and if (U,ϕ) is a chart of M at x, then $\phi(U \cap S) = \phi(U) \cap E_1$. Suppose that (1.2) holds. Then (1.4) and $\phi(x) \in E_1$ imply

$$D(\phi)_x y + r(h) \in E_1, \quad \forall h > 0$$

which yields $D(\phi)_x y \in E_1$, that is, $y \in T_x(S)$. Conversely, if $y \in T_x(S)$ then there exists a chart (U,ϕ) of M at $x \in S$ such that $\phi(x) + hD(\phi)_x y \in \phi(U) \cap E_1 = \phi(U \cap S)$ for all sufficiently small $h > 0$. This means that (1.4) holds with $r(h) = 0$ and therefore y is quasi-tangent to S at x. For each $x \in S$ set

$$\tilde{T}_x(S) = \{y \in T_x(M); y \text{ is quasi-tangent to } S \text{ at } x\}. \tag{1.6}$$

<u>Remark 1.1</u> Theorem 1.1 can be restated as follows: *If* S *is a* C^k*-submanifold of* M, *then* $\tilde{T}_x(S) = T_x(S)$.

Let N be another manifold of class C^k with $k \geq 1$. Recall that $g:M \to N$ is a *submersion* at $x \in M$ iff $D(g)_x$ is surjective and its kernel splits.

The following result is important in flow-invariance and optimization problems. It is also interesting by itself.

<u>Theorem 1.2</u> *Let* M *and* N *be differentiable manifolds and let* $g:M \to N$ *be a* C^1*-mapping.*

(1) *If* $S \subset M$ *is a nonempty subset and* $x \in S$ *then* $D(g)_x(\tilde{T}_x(S)) \subset \tilde{T}_{g(x)}(g(S))$.

(2) *If in addition we suppose that* g *is a submersion at* $x \in M$ *and* $W \subset N$ *is a nonempty subset with* $x \in g^{-1}(W)$, *then*

$$\tilde{T}_x(g^{-1}(W)) = D(g)_x^{-1}(\tilde{T}_{g(x)}(W)) \tag{1.7}$$

$$D(g)_x(\tilde{T}_x(g^{-1}(W)) = \tilde{T}_{g(x)}(W). \tag{1.8}$$

<u>Proof</u> Part (1) is a consequence of the first part of Theorem 2.1 in Chapter 3. (2) In view of Part (1) with $S = g^{-1}(W)$ and of surjectivity of $D(g)_x$, the only fact which remains to be proved is the inclusion

$$D(g)_x^{-1}(\tilde{T}_{g(x)}(W)) \subset \tilde{T}_x(g^{-1}(W)). \tag{1.9}$$

Since g is a submersion at x we may assume (on the basis of the inverse function theorem) that there exist two Banach spaces E' and E" such that $M = E' \times E"$, $N = E'$ and $g:E' \times E" \to E'$ is a projection on the first factor. In other words, if $x = (x_1,x_2) \in M$ and $u = (u_1,u_2) \in T_x(M) = E' \times E"$, then $g(x) = x_1$ and $D(g)_x(u) = u_1$. Now let v be quasi-tangent to $W \subset E'$ at x_1, where $x = (x_1,x_2) \in g^{-1}(W)$. Finally, let $u = (u_1,u_2)$ be such that $D(g)_x(u)=v$. This means that $v = u_1$ and there exists $r(h) \in E'$ with the properties $r(h) \to 0$ as $h \downarrow 0$, $x_1 + h(v + r(h)) \in W$, $\forall h > 0$ (see Lemma 2.1 in Chapter 3). Then it follows that $x + h(u + (r(h),0)) \in g^{-1}(W)$, $0 \in E"$, that is, u is quasi-tangent to $g^{-1}(W)$ at x. The proof is complete.

For some applications the following particular case is important. Namely, take $N = R^n$, $W = \{0\} \subset R^n$, $g:M \to R^n$ and $S = g^{-1}(0)$, that is,

$$S = \{x \in M;\ g(x) = 0\}. \tag{1.10}$$

In this case Theorem 1.2 becomes:

Corollary 1.1 *Let* M *be a differentiable manifold and let* $g:M \to R^n$, $n \geq 1$, *be a* C^1*-mapping. If* g *is a submersion at* $x \in g^{-1}(0)$, *then* $y \in T_x(M)$ *is quasi-tangent to* $g^{-1}(0)$ *iff* $D(g)_x y = 0$ (*i.e.*, $\tilde{T}_x(g^{-1}(0)) = D(g)_x{}^{-1}(0)$).

Remark 1.2 Corollary 1.1 can also be regarded as a consequence of Theorem 1.1. Indeed, if the C^1-mapping g is a submersion at $\forall x \in g^{-1}(0)$, it follows that $g^{-1}(0)$ is a submanifold of M and in addition $T_x(g^{-1}(0)) = D(g)_x{}^{-1}(0)$. Combining this well-known fact with Remark 1.1, we get the con-conclusion of the corollary. The next result shows that the conclusion of Corollary 1.1 holds even in the case in which $g^{-1}(0)$ is not a submanifold.

Let A be an open subset of M and let $g:A \to R^n$. Set

$$S_g = \{x \in A;\ g(x) = 0\}. \tag{1.11}$$

On the basis of Corollary 2.1 in Chapter 3 one obtains:

Corollary 1.2 *Suppose that* $g:A \subset M \to R^n$ *is differentiable at* $x \in A$ *and continuous on the open subset* A *of* M. *If* $D(g)_x$ *is surjective, then* $y \in T_x(M)$ *is quasi-tangent to* S_g *at* $x \in S_g$ *if and only if*

$$D(g)_x y = 0, \quad \textit{i.e.}, \quad \operatorname{Ker} D(g)_x = \tilde{T}_x(S_g). \tag{1.12}$$

Other results on this subject are given in the following sections.

§2. Flow invariance of S with respect to a vector field

In what follows, $X:M \to T(M)$ is a locally Lipschitz vector field on the differentiable manifold M.

Definition 2.1 *A vector field* $X:M \to T(M)$ *is said to be quasi-tangent to the nonempty subset* $S \subset M$, *if for each* $x \in S$, $X(x)$ *is quasi-tangent to* S *at* x (*in the sense of Definition* 1.1), *that is, there exists a chart* (U,ϕ) *of* M *at* x *such that:*

$$\lim_{h \downarrow 0} h^{-1}\, d(\phi(x) + hD(\phi)_x X(x);\ \phi(U \cap S)) = 0. \tag{2.1}$$

In view of Proposition 1.1, this definition is consistent.

Let us consider the differential equation corresponding to the vector field X, that is,

$$x'(t) = X(x(t)), \ x(0) = x_0 \in S, \ t \in J_{x_0} \tag{2.2}$$

where J_{x_0} is the maximal interval of the existence of the integral curve x of X, with $0 \in J_{x_0}$.

Similarly to Definition 4.1 in Chapter 3, the subset S of M is said to be flow-invariant with respect to the vector field X (or Equation (2.2)) if any integral curve x of X starting in S remains in S as long as it exists. One expects that the *tangential* condition (2.1) with "$h \to 0$" in place of "$h \downarrow 0$", that is,

$$\lim_{h \to 0} h^{-1} \, d(\phi(x) + hD(\phi)_x \, X(x); \phi(U \cap S)) = 0, \tag{2.3}$$

characterizes the flow-invariance of S with respect to the vector field X. Of course, as in the case of (2.1), the property (2.3) is also independent of the local chart (U,ϕ) of M at $x \in S$. To distinguish between (2.1) and (2.3), we say that "X is quasi-tangent to S in the sense of (2.3)" if for each $x \in S$ there is a chart (U,) of M at $x \in S$ with the property (2.3).

The extension of the Nagumo-Brezis theorem (i.e., Theorem 4.2, Chapter 2) to Banach manifolds is the following:

Theorem 2.1 *Let* M *be a* C^k *-manifold,* $k \geqslant 1$, *and let* S *be a nonempty closed subset of* M. *Then* S *is flow-invariant with respect to the locally Lipschitz vector field* $X : M \to T(M)$ *if and only if* X *is quasi-tangent to* S *in the sense of* (2.3).

Proof Let $x \in S$ and let (U,ϕ) be a chart of M at x. Set

$$X_\phi(x) = D(\phi)_x \, X(x). \tag{2.4}$$

The local representation of (2.2) in the chart (U,ϕ) is

$$u'(t) = X_\phi o \, \phi^{-1}(u(t)), \ u(0) = \phi(x_0), \ t \in J_\phi \tag{2.5}$$

where $(X_\phi o \phi^{-1})(u) = X_\phi(\phi^{-1}(u))$ for $u \in \phi(U)$. Clearly, (2.4) is a differential equation in the Banach space E (the model of M). It is now essential to observe that (2.3) implies that $X \circ \phi^{-1}$ is tangent to $\phi(U \cap S)$ in the sense of (B4) in 4, Chapter 2; that is,

$$\lim_{h \to 0} h^{-1} d(u + hX_\phi \circ \phi^{-1}(u); \phi(U \cap S)) = 0, \quad \forall u \in \phi(U \cap S). \qquad (2.6)$$

The second crucial point is that $\phi(U \cap S)$ is closed in $\phi(U)$. On the basis of Theorem 1.1 in Chapter 3, Condition (2.6) is necessary and sufficient for the existence of a (unique) solution $u = u_\phi$ to (2.5) with $u_\phi(0) = \phi(x_0) \in \phi(U \cap S)$ and $u_\phi(t) \in \phi(U \cap S)$ for all t in the domain J_ϕ of u_ϕ. Set $x = x_\phi = \phi^{-1} \circ u_\phi$ on J_ϕ. Then $x_\phi(t) \in S$ for all $t \in J_\phi$ and (according to the uniqueness of the solution to (2.5)) x_ϕ is actually independent of ϕ in the following sense: if (V,ψ) is another chart of M at x_0 then $x_\phi = x_\psi$ on $J_\phi \cap J_\psi$. By standard arguments, the local solution x to (2.2) can be extended to a maximal interval $J \subset R$ with $x(t) \in S$ for all $t \in J$. In other words, we have proved that (2.3) ensures the flow-invariance of S with respect to the vector field X on M. Conversely, the flow-invariance of S with respect to X implies (2.6), which is equivalent to (2.3). This completes the proof.

If, in addition to the hypotheses of Theorem 2.1, we assume that S is a compact subset of M, then it is well known that the maximal interval J of the solution x to (2.2) is the whole real axis R. The condition $x(t) \in S$ for all $t \in R$ is ensured by (2.3). Therefore the following result holds:

<u>Corollary 2.1</u> *Let* S *be a compact subset of* M *and let* $X:M \to T(M)$ *be a locally Lipschitz vector field. Then for each* $x_0 \in S$ *the corresponding solution* $x = x(t,x_0)$ *to* (2.2) *satisfies:* "$x(t) \in S$ *for all* $t \in R$", *iff* X *is quasi-tangent to* S *in the sense of* (2.3).

As a first application of Theorem 2.1, we give a characterization of the invariance of a closed subset S of M with respect to a one-parameter group of diffeomorphisms $\{G_t; t \in R\}$ of M. In other words, we will characterize the inclusion

$$G_t(S) \subset S, \quad \forall t \in R. \qquad (2.7)$$

<u>Corollary 2.2</u> *Let* $G_t:M \to M$ *be a one-parameter group of* C^p*-diffeomorphisms*

of the C^k*-manifold* M $(k \geq 2,\ 1 \leq p \leq k-1)$ *and let* S *be a closed subset of* M. *Then the property* (2.7) *is equivalent to: "For every* $x \in S$, *the vector* $X(x) = \frac{d}{dt} G(t,x)|_{t=0}$ *is quasi-tangent to* S *at* x, *in the sense of* (2.3)". (*Here* $G(t,x) = G_t(x)$, *so* $G(0,x) = x$, $\forall x \in M$.)

Obviously, for the proof one applies Theorem 2.1.

Another theoretical application of Theorem 2.1 is the definition of the Lie derivative of tensor fields on S, as shown by the following simple example. Let X be a locally Lipschitz vector field (on M), which is quasi-tangent to the closed subset $S \subset M$, in the sense of (2.3). Denote by G_t the one-parameter group of X. Then (2.7) holds. If Y is the restriction of X to S and $f:S \to R$ is a differentiable function, set

$$(L_Y^f)_x = \lim_{t \to 0} \frac{1}{t} (f(G(t,x)) - f(x)), \quad \forall x \in S. \tag{2.8}$$

In such a situation we can avoid the consideration of local representatives which are used to define the Lie derivative on subcartesian spaces (see Marshall [1]). In our present case, L_Y^f is the Lie derivative on S, of f, with respect to X.

Other uses of Theorem 2.1 are given in the next section.

§3. Flow invariance of S with respect to a second-order differential equation on a Banach manifold

In addition to the assumptions of previous sections, we assume that M is a C^k-manifold with $k \geq 3$. Let (U,ϕ) be a chart of M at x and let $y \in T_x(M)$ and $z \in T_y(T(M))$ with $D(\pi)\, z = y$. For simplicity, the following notation will be adopted:

$$y_\phi = D(\phi)_x y,\ z_{\tilde\phi} = D(\tilde\phi)_y\, z = (z^1_{\tilde\phi}, z^2_{\tilde\phi}),\ z^2_{\tilde\phi} = pr_2\, z_{\tilde\phi} \tag{3.1}$$

where $\tilde\phi$ is defined by (1.1). Clearly,

$$y_\phi \in E,\ z_{\tilde\phi} \in E \times E \text{ and } z^1_{\tilde\phi} = D(\phi)_x\, y = y_\phi. \tag{3.2}$$

If (V,ψ) is another chart at x, then

$$\begin{aligned} z_{\tilde\psi} = D(\tilde\psi\tilde\phi^{-1})_{\tilde\phi(y)}\, z_{\tilde\phi} = (D(\psi\phi^{-1})_{\phi(x)} z^1_{\tilde\phi},\ D^2(\psi\phi^{-1})_{\phi(x)}(y_\phi, z^1_{\tilde\phi}) \\ + D(\psi\phi^{-1})_{\phi(x)} z^2_{\tilde\phi}). \end{aligned} \tag{3.3}$$

Consequently,

$$pr_2\, z_{\tilde{\psi}} = D^2(\psi\phi^{-1})_{\phi(x)}(y_\phi, y_\phi) + D(\psi\phi^{-1})_{\phi(x)}\, pr_2\, z_{\tilde{\phi}}. \tag{3.4}$$

In the sequel, the crucial fact is given by:

<u>Lemma 3.1</u> *Let* $y \in T_x(M)$ *and* $z \in T_y(T(M))$ *with* $D(\pi)_y\, z = y$. *If there is a chart* (U,ϕ) *of* M *at* $x \in S \subset M$ *such that*

$$\lim_{h\downarrow 0} h^{-2}\, d(\phi(x) + hy_\phi + \frac{h^2}{2}\, pr_2\, z_{\tilde{\phi}};\ \phi(U \cap S)) = 0, \tag{3.5}$$

then (3.5) *holds in any other chart* (V,ψ) *at* x.

<u>Proof</u> It is easily seen (see also (3.9)' in Chapter 2) that (3.5) is equivalent to the existence of a function $r: R_+ \to E$, with $r(h) \to 0$ as $h \quad 0$ such that

$$b_h \equiv \phi(x) + hy_\phi + \frac{h^2}{2}(pr_2\, z_{\tilde{\phi}} + r(h)) \in \phi(U \cap S). \tag{3.6}$$

If (V,ψ) is another chart at x, it follows that $\psi\phi^{-1}(b_h) \in \psi(U \cap V \cap S)$ for all sufficiently small $h > 0$. According to the Taylor formula, we have

$$\begin{aligned}\psi\phi^{-1}(b_h) &= \psi(x) + D(\psi\phi^{-1})_{\phi(x)}(b_n - \phi(x)) \\ &\quad + \frac{1}{2} D^2(\psi\phi^{-1})_{\phi(x)}(b_h - \phi(x),\ b_h - \phi(x)) + \bar{r}(b_h - \phi(x))\end{aligned} \tag{3.7}$$

where $\bar{r}(b_h - \phi(x))/h^2 \to 0$ as $h \downarrow 0$. We now have to combine (3.4), (3.7) and $y_\psi = D(\psi\phi^{-1})_{\phi(x)}\, y_\phi$ to get

$$\psi\phi^{-1}(b_h) = \psi(x) + hy_\psi + \frac{h^2}{2}\, pr_2 z_{\tilde{\psi}} + h^2\, r_1(h) \in \psi(U \cap V \cap S) \tag{3.8}$$

with $r_1(h) \to 0$ as $h \downarrow 0$. This means that (3.5) holds with (V,ψ) in place of (U,ϕ), which concludes the proof.

In the case in which S is a submanifold, we have:

<u>Lemma 3.2</u> *Let* S *be a* C^k*-submanifold of* M *and let* $x \in M$, $y \in T_x(M)$ *and* $z \in T_y(T(M))$ *with* $D(\pi)_y z = y$. *Then the following two properties are equivalent:*

(1) *There is a chart* (U,ϕ) *of* M *at* x *such that* (3.5) *holds;*

(2) $x \in S$, $y \in T_x(S)$, $z \in T_y(T(S))$.

Proof Suppose that (3.5) is satisfied. Then (3.6) yields $x \in S$. In addition, because S is a submanifold of M, there is a closed linear subspace E_1 of E with the property that $\phi(U \cap S) = \phi(U) \cap E_1$ (see the proof of Theorem 1.1). Using (3.6) we get

$$y_\phi + \frac{h}{2}(pr_2\, z_{\tilde{\phi}} + r(h)) \in E_1, \quad \forall h > 0. \tag{3.9}$$

First of all (3.9) implies $y_\phi \in E_1$ (that is, $y \in T_x(S)$), and then $pr_2\, z_{\tilde{\phi}} \in E_1$. Since $z^1_{\tilde{\phi}} = y_\phi$, we now have $z_{\tilde{\phi}} = (z^1_{\tilde{\phi}}, z^2_{\tilde{\phi}}) \in E_1 \times E_1$ and therefore $z \in T_y(T(S))$. Conversely, if (2) holds, then using some of the above arguments we conclude that

$$\phi(x) + h\, y_\phi + \frac{h^2}{2}\, pr_2\, z_{\tilde{\phi}} \in \phi(U \cap S)$$

for all sufficiently small $h > 0$. This trivially implies (1) and thus the proof is complete.

We now proceed to the main problems of this section. Recall that a second-order differential equation over M is a vector field $Y:T(M) \to T(T(M))$ (of class C^{k-2}) with the property:

$$D(\pi)_y Y(y) = y, \quad \forall y \in T(M) \tag{3.10}$$

where π denotes the canonical projection of T(M) on M.

It is easy to check that Y is a second-order differential equation over M if and only if each integral curve β of Y is equal to the canonical lifting of $\pi\circ\beta$, that is

$$(\pi\circ\beta)'(t) = \beta(t), \quad \forall t \in J_\beta \tag{3.11}$$

where J_β is the domain of β. In other words, if β satisfies

$$\beta'(t) = Y(\beta(t)), \quad \forall t \in J_\beta, \tag{3.12}$$

then it also satisfies (3.11).

In the theory of orbital motion the following condition is important:

$$\pi(\beta(t)) \in S, \quad \forall t \in J_\beta \tag{3.13}$$

where S is a nonempty subset of M. For the characterization of (3.13) we need the subset

$$M_S = \{y \in T(M);\ \text{there is a chart } (U,\phi) \text{ of } M \text{ at } \pi(y) \text{ such that (3.14)' holds}\}, \tag{3.14}$$

$$\lim_{h \downarrow 0} h^{-2}\, d(\phi(\pi(y)) + hy_\phi + \frac{h^2}{2}\, pr_2 y_{\tilde{\phi}}(y);\ \phi(U \cap S)) = 0. \tag{3.14)'$$

Here y_ϕ and $y_{\tilde{\phi}}$ are defined as in (3.1), that is,

$$y_{\tilde{\phi}}(x) = D(\tilde{\phi})_y\, Y(y),\ y_\phi = D(\phi)_{\pi(y)} y = pr_1\, Y_{\tilde{\phi}}(y). \tag{3.15}$$

It follows from Lemma 3.1 (with $x = \pi(y)$ and $z = Y(y)$) that the definition of M_S is consistent.

For the definition of the flow invariance of S with respect to Y, the following result plays a crucial role:

Proposition 3.1 (1) *Let* S *be a subset of* M. *Then each* $y \in M_S$ *is quasi-tangent to* S *at* $\pi(y)$.

(2) *If* S *is closed then* $\pi(M_S) \subset S$.

(3) *Let* $S \subset M$ *be closed and let* $\beta : J \to T(M)$ *be an integral curve of the differential equation* $Y : T(M) \to T(T(M))$ (*i.e.*, (3.12) *holds*). *Then* (i) *and* (ii) *below are equivalent:*

(i) $\Pi(\beta(t)) \in S, \ \forall t \in J$; (ii) $\beta(t) \in M_S, \ \forall t \in J$.

Proof (1) This follows from the nonexpansivity of the distance function (see (2.3)', Chapter 2). To prove Property (2), take $y \in M_S$ and set $x = \Pi(y)$, $z = Y(y)$. Then (3.14)' is actually (3.6), which yields $\phi(\pi(y)) \in \phi(U \cap S)$ and hence $\pi(y) \in S$. Let us prove (3). The implication "(ii) ⇒ (i)" is a consequence of (2). Consequently, suppose that (i) holds and let $t \in J$. Keeping t fixed, we have $t + h \in J$ for all sufficiently small $h > 0$ and therefore $\pi(\beta(t+h)) \in S$. To prove that $\beta(t) \in M_S$, we first observe that for each chart (U,ϕ) of M at $\pi(\beta(t))$ one has (with $x = \pi(\beta(t))$)

$$(\beta(t))_\phi = D(\phi)_x \beta(t) = (\phi\pi\beta)'(t)$$

$$Y_{\tilde{\phi}}(\beta(t)) = (\tilde{\phi}\beta)'(t) = ((\phi\pi\beta)'(t), (\phi\pi\beta)''(t))$$

(see (3.15)). Accordingly, for all sufficiently small $h > 0$,

$$h^{-2}\, d(\phi(x) + h(\beta(t))_{\phi} + \frac{h^2}{2} pr_2\, Y_{\tilde{\phi}}(\beta(t));\ \phi(U \cap S))$$

$$= h^{-2}\ \|\phi(\pi(\beta(t))) + h(\phi\pi\beta)'(t) + \frac{h^2}{2} (\phi\pi\beta)''(t) - \phi(\pi(\beta(t+h)))\|. \qquad (3.16)$$

Passing to the limit (for $h \downarrow 0$) in (3.16), it follows that $\beta(t) \in M_S$, which completes the proof.

On the basis of Proposition 3.1 the following definition is now quite reasonable:

Definition 3.1 *The subset* S *of* M *is said to be a flow-invariant set with respect to the second-order differential equation* $Y:T(M) \to T(T(M))$, *if* M_S *is nonempty and if for each integral curve* $\beta: J \to T(M)$ *of* Y *with* $\beta(0) \in M_S$ *we have* $\pi(\beta(t)) \in S$, *for all* $t \in J$, $t \geqslant 0$.

On the basis of Proposition 3.1, Definition 3.1 can be restated in the following standard form:

Definition 3.1' $S \subset M$ *is flow-invariant with respect to the second-order differential equation* $Y:T(M) \to T(T(M))$ *if every integral curve* β *of* Y *starting in* M_S *at* $t = 0$, *remains in* M_S *as long as it exists (to the right of zero).*

Finally, we can give the main result on the flow-invariance of S with respect to Y:

Theorem 3.1 *Let* $Y:T(M) \to T(T(M))$ *be a locally Lipschitz second-order differential equation over* M. *Suppose that both* S *and* M_S *are nonempty and closed. Then* S *is a flow-invariant set with respect to* Y *if and only if* Y *is quasi-tangent to* M_S, *that is, for each* $y \in M_S$ *there is a chart* $(\tilde{U},\tilde{\phi})$ *of* $T(M)$ *at such that*

$$\lim_{h\downarrow 0} h^{-1}\, d(\tilde{\phi}(y) + h\, D(\tilde{\phi})_y Y(y);\ \tilde{\phi}(U \cap M_S)) = 0. \qquad (3.17)$$

Proof One applies Theorem 2.1 with $T(M)$, Y and M_S in place of M, X and S respectively. Clearly, if in Theorem 2.1 we replace (2.3) by (2.1), we get only the invariance of S to the right of zero.

§4. Flow-invariance of a submanifold and applications to orbital motion

In this section we present some consequences of the results from the previous section, in the case in which S is a submanifold of the C^k-manifold M, with $k \geq 3$.

The main tool is the transversality condition. The first result we need in the sequel is:

Lemma 4.1 *Let* $Y:T(M) \to T(T(M))$ *be a second-order differential equation over* M *and let* S *be a submanifold of* M. *Then the subset* M_S *defined by* (3.14) *is given by*

$$M_S = Y^{-1}(T(T(S))). \tag{4.1}$$

Proof It follows from Lemma 3.2 with $y \in T(M)$, $x = \pi(y)$ and $z = Y(y)$.

4.1 Differentiability preserves transversality

Let M and N be manifolds of class C^k, $k \geq 3$, modelled on Banach spaces E and F respectively. Let us recall the concept of transversality of a function $g:M \to N$ over a submanifold W of N (see Lang [1]).

Definition 4.1 *The* C^k*-morphism* $g:M \to N$ *is said to be transversal over the submanifold* W *of* N *if, for each* $x \in g^{-1}(W)$,

$$T_{g(x)}(N) = D(g)_x(T_x(M)) + T_{g(x)}(W) \tag{4.2}$$

and $D(g)_x^{-1}(T_{g(x)}(W))$ *splits in* $T_x(M)$.

This definition is equivalent to the fact that, for $x \in g^{-1}(W)$, the composite map

$$T_x(M) \xrightarrow{D(g)_x} T_{g(x)}(N) \to T_{g(x)}(N)/T_{g(x)}(W)$$

is surjective and its kernel splits in $T_x(M)$.

We shall use the following well-known result:

Theorem 4.1 *If* $g:M \to N$ *is transversal over the submanifold* W *of* N *then* $g^{-1}(W)$ *is a submanifold of* M *and*

$$T_x(g^{-1}(W)) = D(g)_x^{-1}(T_{g(x)}(W)), \quad \forall x \in g^{-1}(W)$$

i.e.,

$$T(g^{-1}(W)) = (D(g))^{-1}(T(W)). \tag{4.3}$$

In what follows, the following result will be essentially used. It is also interesting in itself.

Lemma 4.2 *If the* C^2*-map* $g:M \to N$ *is transversal over the submanifold* W *of* N, *then* $D(g):T(M) \to T(N)$ *is transversal over the submanifold* $T(W)$ *of* $T(N)$. *(In short, we could say: the differentiability preserves transversality.)*

Proof Let $x \in M$ be such that $y = g(x) \in W$. It is known that the transversality of g over W is equivalent to the existence of local charts (U,ϕ) at x and (V,ψ) at y with the properties: $\phi(U) = U_1 \times V_2$, $\psi(V) = V_1 \times V_2$, $g(U) \subset V$, $\phi(x) = (0,0)$, $\psi(y) = (0,0)$, $\psi(W \cap V) = V_1 \times \{0\}$, where $U_1 \subset E_1$, $V_i \subset F_i$, $i = 1,2$, and E_1, F_1 are Banach spaces such that M and N are modelled on $E_1 \times F_2$ and $F_1 \times F_2$ respectively. In addition, the local representation $g_{\phi\psi} = \psi g \phi^{-1}:U_1 \times V_2 \to V_1 \times V_2$ of g is given by (see Lang [1], p. 22)

$$g_{\phi\psi}(x_1,x_2) = (f(x_1,x_2),x_2), \forall (x_1,x_2) \in U_1 \times V_2.$$

Consequently

$$\tilde{\psi}D(g)\tilde{\phi}^{-1}(x_1,x_2,u_1,v_2) = (f(x_1,x_2),x_2,f'_{x_1}u_1 + f'_{x_2}v_2,v_2)$$

for all $(u_1,u_2) \in E_1 \times F_2$, with $f'_z = \partial f/\partial z$.

Now let $I_1:F_1 \times F_2 \times F_1 \times F_2 \to F_1 \times F_1 \times F_2 \times F_2$ and $I_2:E_1 \times F_2 \times E_1 \times F_2 \to E_1 \times E_1 \times F_2 \times F_2$ be the isomorphisms

$$I_j(a,b,c,d) = (a,c,b,d), \; j = 1,2.$$

If we set $\tilde{\psi}_1 = I_1\tilde{\psi}$ and $\tilde{\phi}_1 = I_2^{-1}\tilde{\phi}$, then we have

$$\tilde{\psi}_1 D(g)\tilde{\phi}_1^{-1}(x_1,u_1,x_2,v_2) = (f(x_1,x_2),f'_{x_1}u_1 + f'_{x_2}v_2,x_2,v_2) \tag{4.4}$$

for all $(x_1,u_1,x_2,v_2) \in U_1 \times E_1 \times V_2 \times F_2$. This is equivalent to the transversality of $D(g)$ at y over $T(W)$, which completes the proof.

Remark 4.1 The converse assertion of Lemma 4.2 is also true (see §7).

4.2 Flow invariance of S and applications

We are now prepared to prove:

Theorem 4.1 *Suppose that the* C^2*-morphism* $g:M \to N$ *is transversal over the submanifold* W *of* N. *Then for* $S = g^{-1}(W)$ *the subset* M_S *given by* (4.1) *becomes*

$$M_S = (D^2(g)\circ Y)^{-1}(T(T(W)) \tag{4.5}$$

Proof In view of (4.3), we have

$$T(S) = T(g^{-1}(W)) = (D(g))^{-1}(T(W)).$$

Using once again (4.3) with $D(g)$ and $T(W)$ in place of g and W respectively (this is possible on the basis of Lemma 4.2), it follows that

$$T(T(S)) = (D(D(g)))^{-1}(T(T(W))).$$

Consequently (4.1) implies (4.5).

For application to orbital motion the following particular case is the most important:

$$N = R^n,\ g:M \to R^n,\ W = \{0\} \subset R^n$$

and (4.6)

$$S = g^{-1}(0) = \{x \in M;\ g(x) = 0\}.$$

In this case Theorem 4.1 yields (for $W = \{0\}$):

Corollary 4.1 *Let* $g:M \to R^n$ *be a* C^2*-function. Suppose that* $0 \in R^n$ *is a regular value of* g. *Then*

$$M_{g^{-1}(0)} = \{y \in T(M);\ D^2(g)(Y(y)) = 0\}. \tag{4.7}$$

Proof In this case, g is transversal over $\{0\} \subset R^n$, and therefore (4.5) becomes simply (4.7).

We can now prove the main result of this section, that is:

Theorem 4.2 *Let* $g:M \to R^n$ *be a* C^3*-mapping and let* $0 \in R^n$ *be a regular value of* g. *Suppose that* $Y:T(M) \to T(T(M))$ *is a second-order differential equation of class* C^1 *and that for each* $y \in M_{g^{-1}(0)}$, *the mapping*

$$z \to (D(g)_{\pi(y)}(D(\pi)_y z),\ pr_2 D^2(g)(y)(z),\ pr_4 D^3(g)(Y(y))(D(Y)_y z)) \qquad (4.8)$$

from $T_y(T(M))$ *into* R^{3n} *is surjective. Then* $g^{-1}(0)$ *is flow-invariant with respect to* Y *if and only if*

$$D(D^2(g))_{Y(y)}(D(Y)_y(Y(y))) = 0, \quad \forall y \in M_{g^{-1}(0)} \qquad (4.9)$$

where pr_2 *and* pr_4 *denote the projections onto the second and fourth factor, respectively.*

Proof Let us observe that $M_{g^{-1}(0)}$ given by (4.7) can be written in the form

$$M_{g^{-1}(0)} = \{y \in T(M);\ K(y) = 0\} \qquad (4.10)$$

with

$$K(y) = (g(\pi(y)),\ pr_2 D(g)(y), pr_4 D^2(g)(Y(y))). \qquad (4.11)$$

Obviously, Hypothesis (4.8) means that 0 is a regular value of K. Therefore, by Corollary 1.2, $Y(y)$ is quasi-tangent to $M_{g^{-1}(0)}$ at $y \in K^{-1}(0) = M_{g^{-1}(0)}$ if and only if

$$D(K)_y\ Y(y) = 0 \qquad (4.12)$$

which is equivalent to (4.9). Now the conclusion of the theorem follows from that of Theorem 3.1. The proof is complete.

We now show that the results of this section cover most of the results in Chapter 3. Indeed, let us see the local version of the previous theorems: we assume that M is a Banach space E. In this case, the vectors $y \in T_x(M)$ and $Y(y) \in T_y(T(M))$ are respectively identified with

$$(x,y),\ (x,y,y,f(x,y));\ (x,y,f(x,y) \in E). \qquad (4.13)$$

As in the previous chapter, denote by $\dot{g}(x)(y)$ the Fréchet derivative of

$g:E \to R^n$ at x in the direct y E. We now have (for all $x,y,u,v \in E$)

$$D(g)(x,y) = (g(x),\dot{g}(x)(y));$$
$$D^2(g)(x,y)(u,v) = (g(x),\dot{g}(x)(y),\dot{g}(x)(u),\ddot{g}(x)(y)(u) + \dot{g}(x)(v)) \tag{4.14}$$

$$\begin{aligned} D^3(g)(x,y)(u,v)(\bar{x},\bar{y},\bar{u},\bar{v}) &= (g(x),\dot{g}(x)(y),\dot{g}(x)(u), \\ &\ddot{g}(x)(y)(u) + \dot{g}(x)(v),\dot{g}(x)(\bar{x}),\ddot{g}(x)(\bar{x})(y) + \dot{g}(x)(\bar{y}), \\ &\ddot{g}(x)(\bar{x})(u) + \dot{g}(x)(\bar{u}),\ \dddot{g}(x)(\bar{x})(y)(u) + \ddot{g}(x)(\bar{y})(u) \\ &\quad + \ddot{g}(x)(y)(\bar{u}) + \ddot{g}(x)(\bar{x})(v) + \dot{g}(x)(\bar{v})). \end{aligned} \tag{4.15}$$

Since $Y(y) = (x,y,y,f(x,y))$ (as indicated in (4.13), it follows that

$$\begin{aligned} D(Y)(x,y)(v_1,v_2) = (x,y,y,f(x,y),v_1,v_2,v_2, \\ \dot{f}_x(x,y)(v_1) + \dot{f}_y(x,y)(v_2)) \end{aligned} \tag{4.16}$$

where $v_1,v_2 \in E$ and $\dot{f}_x(x,y)(v_1)$ is the derivative of f with respect to x evaluated at v_1.

The second-order differential equation $Y:T(M) \to T(T(M))$ is represented by its principal part $(y,f(x,y))$ with

$$y = x' = \frac{dx}{dt},\ y' = f(x,y)$$

and therefore

$$\frac{d^2x}{dt^2} = f(x,y) = f(x,x'). \tag{4.17}$$

According to (4.13) and (4.14), Corollary 4.1 becomes:

<u>Corollary 4.2</u> *If* $0 \in R^n$ *is a regular value of the* C^2*-function* $g:E \to R^n$, *then*

$$\begin{aligned} M_{g^{-1}(0)} &= \{(x,y) \in E \times E;\ \ddot{g}(x)(y)(y) + \dot{g}(x)(f(x,y)) = 0,\ g(x) = 0, \\ &\qquad \dot{g}(x)(y) = 0\} \\ &= \{(x,y) \in E \times E;\ \lim_{h \downarrow 0} h^{-2}d(x+hy + \frac{h^2}{2}f(x,y);g^{-1}(0)) = 0 \end{aligned} \tag{4.18}$$

Proof We replace in (3.14)': $x = \phi(\pi(y))$, $\phi(U \cap S) = g^{-1}(0)$,

$$y_{\tilde{\phi}}(y) = (y,f(x,y)),\ pr_2 Y_{\tilde{\phi}}(y) = f(x,y), y_{\phi} = y. \tag{4.19}$$

The local version of Theorem 4.2 is given by:

Theorem 4.3 *Suppose that the following conditions are fulfilled:*

(1) $0 \in R^n$ *is a regular value of the* C^3*-function* g *from the Banach space* E *into* R^n *and* $M_{g^{-1}(0)}$ *is nonempty.*

(2) $f: E \times E \to E$ *is of class* C^1 *and for each* $(x,y) \in M_{g^{-1}(0)}$ *the mapping*

$$v \to (\dot{g}(x)(v), [\ddot{g}(x)(y) + 2^{-1}\dot{g}(x)(\dot{f}_y(x,y))](v)) \tag{4.20}$$

from E *into* R^{2n} *is surjective. Then* $S = g^{-1}(0)$ *is flow-invariant with respect to Equation* (4.17) (*in the sense of Definition* 3.1, *or equivalently of Definition* 1.2 *in Chapter* 3) *if and only if*

$$\begin{aligned}\dddot{g}(x)(y)(y)(y) + 3\ddot{g}(x)(y)(f(x,y)) + \dot{g}(x)(\dot{f}_x(x,y)(y))\\ + \dot{g}(x)(\dot{f}_y(x,y)(f(x,y)) = 0\end{aligned} \tag{4.21}$$

for all x *and* y *in* E *with the properties*

$$g(x) = 0,\ \dot{g}(x)(y) = 0,\ \ddot{g}(x)(y)(y) + \dot{g}(x)(f(x,y)) = 0$$

(*i.e.*, $(x,y) \in M_{g^{-1}(0)}$).

Proof We have to prove that Hypothesis (4.20) implies (in our case here) the surjectivity of the mapping (4.8). On the basis of (4.14) and (4.15), the mapping (4.8) is now

$$\begin{aligned}(u,v) \to (\dot{g}(x)(u), \ddot{g}(x)(y)(u) + \dot{g}(x)(v), \dddot{g}(x)(y)(y)(u) + 2\ddot{g}(x)(y)(v)\\ + \ddot{g}(x)(u)(f(x,y)) + \dot{g}(x)(\dot{f}_x(x,y)(u)) + \dot{g}(x)(\dot{f}_y(x,y)(v)))\end{aligned} \tag{4.22}$$

Indeed, $\pi(x,y) = x$ and therefore $D(\pi)(x,y)(u,v) = (x,u)$. Moreover, by (4.16), the last component in (4.8) is the last component of

$$D^3(g)(x,y)(y,f(x,y))(u,v,v,\dot{f}_x(x,y)(u) + \dot{f}_y(x,y)(v)) \tag{4.23}$$

which is given by (4.15). It is now clear that (4.22) is the local version of (4.8). We now prove that the surjectivity of (4.20) implies the surjectivity of (4.22). For this purpose, let $y_1, y_2, y_3 \in R^n$. We show that for each $(x,y) \in M_{g^{-1}(0)}$ there are $u, v \in E$ such that

$$\dot{g}(x)(u) = y_1; \ \ddot{g}(x)(y)(u) + \dot{g}(x)(v) = y_2$$

$$\dddot{g}(x)(y)(y)(u) + \ddot{g}(x)(u)(f(x,y)) + \dot{g}(x)(\dot{f}_x(x,y))(u)) \qquad (4.24)$$

$$+ 2\ddot{g}(x)(y)(v) + \dot{g}(x)(\dot{f}_y(x,y)(v)) = y_3.$$

By hypothesis, $\dot{g}(x):E \to R^n$ is surjective (since in (4.24) $(x,y) \in M_{g^{-1}(0)}$). Accordingly, there is $u \in E$ which satisfies the first equation of (4.24). The existence of v satisfying the other two equations of (4.24) is a consequence of the surjectivity of the mapping given by(4.20). Finally, it is easily seen that (4.9) is now equivalent to (4.21). Indeed, (4.9) means (in this case)

$$D^3(g)(x,y)(y,f(x,y))(y,f(x,y),f(x,y),\dot{f}_x(x,y)(y) + \dot{f}_y(x,y)(f(x,y)) = 0 \qquad (4.24)$$

for all $(x,y) \in M_{g^{-1}(0)}$ given by (4.18).

It is now clear that, taking into account (4.15), (4.24), is simply (4.21). This completes the proof.

Remark 4.2 In the case in which f(x,y) = f(x) (i.e., the force f is independent of speed), then $\dot{f}_y = 0$, and therefore from Theorem 4.3 one obtains Theorem 1.6 in Chapter 3. Some particular forms of the invariance condition (3.17) of S with respect to Y are given by (4.9) and (4.21). Other forms of (3.17) are given in the next section. It is interesting to note that the equivalence of (3.17) with each of (4.9) and (4.21) has been proved by the transversality approach (i.e., via Theorem 4.1). In Chapter 3, such an equivalence was proved via the fixed point theorem of Brower.

§5. The transversality approach to flow invariance, without surjectivity

It is the purpose of this section to give a result similar to Theorem 4.2, without the surjectivity assumption (4.8). We shall use only the transversality approach given by Theorem 4.1 and Lemma 4.2.

Throughout this section, S is a submanifold (modelled on E') of the C^k-manifold M, $k \geq 3$, modelled on the Banach space E. Consequently $E = E' \oplus E''$. Denote by $p:T(T(M)) \to T(M)$ and $\pi:T(M) \to M$ the canonical projections, and define $T_1(T(S)) = \{z \in T(T(M));\ p(z) \in T(S)$ and there exists a chart (U,ϕ) of M at $\pi(p(z))$ such that $\phi(U \cap S) = \phi(U) \cap E'$ and

$$D(D(\phi))_{p(z)}.z \in E' \times E\}. \tag{5.1}$$

It is easily seen that the following result holds (see also (3.3)):

Lemma 5.1 *The subset* $T_1(T(S))$ *is a submanifold of* $T(T(M))$ *modelled on* $E' \times E' \times E' \times E$. *In addition, every second-order differential equation* $Y:T(M) \to T(T(M))$ *maps* $T(S)$ *into* $T_1(T(S))$.

The main result of this section is the theorem below:

Theorem 5.1 *Let* S *be a closed submanifold of* M *and let* $Y:T(M) \to T(T(M))$ *be a second-order differential equation of class* C^1. *Suppose that the restriction* $Y|_{T(S)}:T(S) \to T_1(T(S))$ *is transversal over the submanifold* $T(T(S))$ *of* $T_1(T(S))$, *that is, for* $y \in Y^{-1}(T(T(S)))$ *we have*

$$D(Y|_{T(S)})_y(T_y(T(S))) + T_{Y(y)}(T(T(S))) = T_{Y(y)}(T_1(T(S))). \tag{5.2}$$

Then S *is flow-invariant with respect to* Y *if and only if*

$$D(Y)_y.Y(y) \in T_{Y(y)}(T(T(S))), \quad \forall y \in Y^{-1}(T(T(S))). \tag{5.3}$$

Proof By Lemma 4.1 we know that $M_S = Y^{-1}(T(T(S)))$. On the basis of Theorem 4.1, Hypothesis (5.2) assures the fact that M_S is a submanifold of T(S) and that

$$T_y(M_S) = D(Y)_y^{-1}(T_{Y(y)}(T(T(S))), \quad \forall y \in M_S. \tag{5.4}$$

According to Theorems 3.1 and 1.1, S is flow-invariant with respect to Y if and only if for every $y \in M_S$, Y(y) is quasi-tangent to M_S at y, that is, $Y(y) \in T_y(M_S)$. This fact is equivalent to (5.3). Of course we have also used the property that $Y(y) \in T_y(T(S))$ for all $y \in M_S$. The proof is complete.

An immediate consequence of this theorem is:

<u>Corollary 5.1</u> *Let* $g:M \to N$ *be a* C^k*-mapping which is transversal over the closed submanifold* W *of* N. *Let also* $Y:T(M) \to T(T(M))$ *be a second-order differential equation of class* C^1 *such that the restriction*

$$Y|_{D(g)^{-1}(T(W))} : D(g)^{-1}(T(W)) \to T_1(D(g)^{-1}(T(W)))$$

is transversal over $D(D(g))^{-1}(T(T(W)))$. *Then* $g^{-1}(W)$ *is flow-invariant with respect to* Y *if and only if*

$$D(D^2(g))_{Y(y)}(D(Y)_y.Y(y)) \in T_{D^2(g)(Y(y))}(T(T(W))) \qquad (5.5)$$

for all $y \in (D^2(g)\circ Y)^{-1}(T(T(W)))$.

<u>Proof</u> The hypotheses imply that $S = g^{-1}(W)$ is a submanifold of M and that $M_S = (D^2(g)\circ Y)^{-1}(T(T(W)))$ is a submanifold of $D(g)^{-1}(T(W))$. By the same combination of Theorem 4.1 and Lemma 4.2, we have

$$\begin{aligned} T_{Y(y)}(T(T(S))) &= T_{Y(y)}(D^2(g)^{-1}(T(T(W))) \\ &= D(D^2(g))^{-1}_{Y(y)}(T_{D^2(g)(Y(y))}(T(T(W))), \quad \forall y \in M_S. \end{aligned} \qquad (5.6)$$

In view of (5.6), we see that (5.3) is equivalent to (5.5) and thus the proof is complete.

A significant particular case of Corollary 5.1 is the following:

<u>Corollary 5.2</u> *Let* $0 \in R^n$ *be a regular value of the* C^k*-mapping* $g:M \to R^n$ *and let* $Y:T(M) \to T(T(M))$ *be a second-order differential equation of class* C^1. *Suppose that*

$$\begin{aligned} D(Y)_y(D(D(g))_y^{-1}(0) + D(D^2(g))^{-1}_{Y(y)}(0) &= T_{Y(y)}(T_1(D(g)^{-1}(0))), \\ &\forall y \in (D^2(g)\circ Y)^{-1}(0). \end{aligned} \qquad (5.7)$$

Then $S = g^{-1}(0)$ *is flow-invariant with respect to* Y *iff*

$$D(D^2(g))_{Y(y)}(D(Y)_y.Y(y)) = 0, \quad \forall y \in (D^2(g)\circ Y)^{-1}(0). \qquad (5.8)$$

<u>Proof</u> Set $W = \{0\} \subset R^n$. Then (4.5) becomes $M_{g^{-1}(0)} = (D^2(g)\circ Y)^{-1}(0)$. The

second remark is that Condition (5.7) means simply the transversality of $Y_{|D(g)^{-1}(0)} : D(g)^{-1}(0) \to T_1(D(g)^{-1}(0))$ over the submanifold $(D^2(g))^{-1}(0)$ of $T_1(T(S)) = T_1(D(g)^{-1}(0))$. Indeed, by (5.6) and (4.3) we have, respectively,

$$T_{Y(y)}((D^2(g))^{-1}(0)) = D(D^2(g))^{-1}_{Y(y)}(0),\ T_y(D(g)^{-1}(0)) = (D(D(g))^{-1}_y(0). \tag{5.9}$$

In view of Corollary 5.1, the proof is complete.

It is interesting to compare the results of Theorem 4.3 and Corollary 5.2. We first examine the surjectivity condition (4.20). Assume for simplicity that $f(x,y) = f(x)$ is independent of y. Then Condition (4.20) means the surjectivity of

$$v \to (\dot{g}(x)(v),\ \ddot{g}(x)(y)(v)),\quad \forall (x,y) \in M_{g^{-1}(0)} \tag{5.10}$$

which acts from E into R^{2n}. This is actually the basic hypothesis of the Theorem 1.6 in Chapter 3. Let us observe that in the case

$$E = R^m,\ m < 2n \tag{5.11}$$

Condition (5.10) is impossible (consequently, we must necessarily have $m \geqslant 2n$). Indeed, if $g: R^m \to R^n$, then the surjectivity of the linear functional given by (5.10) is equivalent to the linear independency of $2n$ linear functionals (which act from R^m into R). Alternatively, $2n$ linear functionals on R^m, with $m < 2n$, are always linear dependent. In other words, if $E = R^m$, then Condition (5.10) implies $m \geqslant 2n$.

To see the restrictions imposed on m and n by Condition (5.7), we need the following addendum to Theorem 4.1:

<u>Remark 5.1</u> If, in addition to the hypotheses of Theorem 4.1, we suppose that M and N are finite dimensional, then the codimension of $g^{-1}(W)$ in M is equal to the codimension of W in N, that is,

$$\dim M - \dim g^{-1}(W) = \dim N - \dim W. \tag{5.12}$$

Now let us assume, as in (5.11), that $\dim M = m$ and $\dim N = n$. Then a necessary condition in order to have (5.2) is given by: $2 \dim S + 4 \dim S \geqslant 3 \dim S + \dim M$, i.e., $3 \dim S \geqslant \dim M$. In the case of Corollary 5.2 we have

$N = R^n$ $W = \{0\}$ and $S = g^{-1}(0)$. Consequently, if M is finite dimensional then (5.12) yields dim S = dim M - n = m - n, which (on the basis of the previous inequality) implies

$$m \geqslant \frac{3}{2} n. \qquad (5.13)$$

In other words, in addition to the surjectivity condition (5.10), the transversality condition (5.7) covers the cases

$$\frac{3}{2} n < m \leqslant 2n. \qquad (5.14)$$

Thus, from the point of view of dimensions the transversality condition (5.7), which covers the cases (5.13), is better than the surjectivity condition (4.20) which covers only the cases $m \geqslant 2n$. The final conclusion is that the situations

$$m \leqslant \frac{3}{2} n \qquad (5.15)$$

remain unsolved by Corollary (5.2), although (5.15) are not excluded by Theorem 3.1.

§6. Quasi-tangent vectors in optimization

The theory of quasi-tangent vectors has also a unifying effect in optimization. This is exactly what we want to point out in this section. We first give a general result (Theorem 6.1) in terms of quasi-tangent vectors and then we show that it contains both the Euler equation from the calculus of variations and the Lagrange multipliers rule. More general results in terms of generalized gradients are also briefly given.

6.1 The differentiable case. Euler equation. Lagrange multipliers

Let M be a C^k-manifold, $k \geqslant 1$, modelled on the Banach space E, and let A be an open subset of M. We shall be concerned with the problem

$$\min \{F(x); x \in S\} = F(x_o), \; x_o \in S \qquad (6.1)$$

where $F: A \to R$ is a function and S is a nonempty (closed in A) subset of A.

A necessary condition in order that (6.1) be fulfilled is given by:

Theorem 6.1 *Suppose that* (6.1) *holds and that* F *is differentiable at* x_o. *Then* $D(F)_{x_o} y = 0$, *for all* $y \in T_x(M)$ *which are quasi-tangent to* S *at* $x_o \in S$ *in the sense of* (1.2) *with* "$\lim_{h\to 0}$" *in place of* "$\lim_{h\downarrow 0}$", *i.e.*,

$$\lim_{h\to 0} h^{-1} d(\phi(x_o) + hD(\phi)_{x_o} y; \phi(U \cap S)) = 0 \tag{6.2}$$

where (U,ϕ) *is a chart of* M *at* x_o.

Proof Let V_{x_o} be a neighbourhood of $\phi(x_o)$. Set $y_\phi = D(\phi)_{x_o} y$. By hypothesis we have $F\circ\phi^{-1}(u) \geqslant F(x)$ for all $u \in V_{x_o} \cap \phi(U \cap S)$, where $V_{x_o} \subset \phi(U)$. Arguing as in the case of (1.2), it is easily seen that (6.2) is equivalent to the existence of a function $r_\phi = r:R \to E$ with the properties: $r(h) \to 0$ as $h \to 0$ and

$$u_h = \phi(x_o) + hy_\phi + hr(h) \in \phi(U \cap S), \quad \forall h \in R. \tag{6.3}$$

For $u = u_h$, the preceding inequality yields

$$F\circ\phi^{-1}(\phi(x_o) + hy_\phi + hr(h)) - F\circ\phi^{-1}(\phi(x_o)) \geqslant 0 \tag{6.4}$$

for all $h \in R$ for which $u_h \in V_{x_o}$ (say $|h| < h_o$). Combining the definition of Fréchet derivative and (6.4) with

$0 < h < h_o$, $-h_o < h < 0$ we get $D(F\circ\phi)_{\phi(x_o)} y_\phi \geqslant 0$, $D(F\circ\phi)_{\phi(x_o)} y_\phi \leqslant 0$,

respectively. This completes the proof.

An immediate consequence of this theorem is the following classical form of the Lagrange multipliers rule:

Corollary 6.1 *Suppose that* $g:M \to R$ *is a continuous function and that* $0 \in R$ *is a regular value of* g. *If*

$$\min \{F(x); x \in g^{-1}(0)\} = F(x_o), \; g(x_o) = 0 \tag{6.5}$$

and if F *is differentiable at* x_o, *then there exists a constant* $k \in R$, $k \neq 0$, *such that*

$$D(F)_{x_0} = k\, D(g)_{x_0} \tag{6.6}$$

Proof Since $0 \in R$ is a regular value of g, it follows that $S = g^{-1}(0)$ is a submanifold of M. In this case, each of (1.2) and (6.2) is equivalent to $y \in T_{x_0}(g^{-1}(0))$ (see the proof of Theorem 1.1). In other words, (1.2) and (6.2) are equivalent in this case. By Corollary 1.2 we now have

$$T_{x_0}(g^{-1}(0)) = \mathrm{Ker}\, D(g)_{x_0} \tag{6.7}$$

where $\mathrm{Ker}\, D(g)_{x_0} = \{y \in T_{x_0}(M);\ D(g)_{x_0} y = 0\}$. Accordingly, the conclusion of Theorem 6.1 becomes

$$\mathrm{Ker}\, D(g)_{x_0} \subset \mathrm{Ker}\, D(F)_{x_0}. \tag{6.8}$$

Clearly, (6.8) is equivalent to (6.6) and thus the proof is complete.

We now proceed to derive the Euler equation from Theorem 6.1. Let $a,b \in R$ with $a < b$ and let $f:M \times T(M) \to R$ be a C^1-function. Set

$$S = \{x \in C^1([a,b];M);\ x(a) = p,\ x(b) = q\} \tag{6.9}$$

where p and q are two given points of M. In other words, S is the set of all C^1-paths $x:[a,b] \to M$ which join p and q. The classical problem in the calculus of variations is to minimize

$$F(x) = \int_a^b f(x(t),x'(t))dt \tag{6.10}$$

over S given by (6.9).

In order to apply Theorem 6.1 we need to characterize the quasi-tangency to S. For this purpose we recall that $C^1([a,b];M)$ admits a structure of differentiable manifold such that for every $x \in C^1([a,b];M)$ there exist a diffeomorphism H and a chart $\phi_x:U_x \to \Gamma^1(x^*T(M))$ of the form $\phi_x(u) = H\circ u$, where $\Gamma^1(x^*T(M))$ is the vector space of all C^1-vector fields along the path x. For this problem we refer to Bourbaki [1]. The following result holds:

Lemma 6.1 *Let* $x \in S$ *with* S *given by* (6.9). *A vector* $y \in T_x(C^1([a,b];M)$ *is quasi-tangent to* S *at* x *in the sense of* (1.2) (*or equivalently of* (6.2)) *if and only if*

$$y_{\phi_x}(a) = y_{\phi_x}(b) = 0 \tag{6.11}$$

where $y_{\phi_x} = D(\phi_x)_x\, y$ *(similarly to* (3.1)).

Proof Suppose that y is quasi-tangent to S at x in the sense of (1.2) with $C^1([a,b];M)$ in place of M. This is equivalent to the existence of $r(h) \to 0$ as $h \downarrow 0$ such that

$$a_h = \phi_x(x) + h(y_{\phi_x} + r(h)) \in \phi_x(U_x \cap S), \quad \forall h > 0. \tag{6.12}$$

In our present case this means that

$$a_h = \phi_x(u_h) = H \circ u_h \text{ for some } u_h \in U_x \cap S. \tag{6.13}$$

We now have

$$(\phi_x(x))(a) = H(x(a)) = a_h(a) = H(u_h(a)) = H(p)$$

which (in view of (6.12)) implies $y_{\phi_x}(a) = 0$. Similarly, $y_{\phi_x}(b) = 0$. In other words, the quasi-tangency of y to S at $x \in S$ implies (6.11). Conversely, (6.11) yields

$$\phi_x(x) + h\, y_{\phi_x} \in \phi_x(U_x \cap S) \tag{6.14}$$

for all $h \in R$ with sufficiently small $|h|$. Obviously, (6.14) implies (6.12) with $C^1([a,b];M)$ in place of M. The proof is complete.

Remark 6.1 From the proof of Lemma 6.1 it follows that, in the case of S given by (6.9), the quasi-tangency (to the right) in the sense of (1.2) coincides with that given by (6.2).

Corollary 6.2 *Suppose that* (6.1) *holds with* F *and* S *given by* (6.10) *and* (6.9). *Then necessarily* x_0 *satisfies the Euler equation*

$$\dot{f}_x(x_0,x_0') - \frac{d}{dt}\dot{f}_{x'}(x_0,x_0') = 0. \tag{6.15}$$

Proof Combining Theorem 6.1 and Lemma 6.1, we get

$$\int_a^b (\dot f_x(x_o,x_o') - \frac{d}{dt}\dot f_{x'}(x_o,x_o'))(D(\phi_{x_o})_{x_o})^{-1}(y_{\phi_{x_o}})dt = 0$$

for all $y_{\phi_{x_o}} \in \Gamma^1(x^*T(M))$ satisfying (6.11). This fact implies (6.15) by standard arguments.

Remark 6.2 In the case in which M is a Banach space E, Lemma 6.1 asserts that $y \in C^1([a,b];E)$ is quasi-tangent to S given by (6.9) if and only if $y(a) = y(b) = 0$. Using the notation (1.6) this means that for each $x \in S$ we have

$$\tilde T_x(S) = \{y \in C^1([a,b];E);\ y(a) = y(b) = 0\} \tag{6.16}$$

Of course in this case the quasi-tangency of y to S at x is given by (2.1) in Chapter 3, i.e.,

$$\lim_{h\downarrow 0} h^{-1}\, d(x + hy;\ S) = 0 \tag{6.17}$$

which is actually (1.2) with $M = C^1([a,b];E)$.

6.2 Lagrange multipliers in the nondifferentiable case. Convexity of Clarke's cone. Generalized gradients

We have already seen that the notion of quasi-tangency to S in the sense of (6.17) is useful both in flow-invariance theory and in optimization problems under differentiability assumptions. Sometimes we are led to consider the problem (6.1) with F locally Lipschitz, only. To be able to treat (6.1) in the nondifferentiable case we need to define a special notion of quasi-tangency as well as the generalized gradients, in the sense of Clarke [1]. In fact, we want to demonstrate that by using a technique from §1 one can extend some results of Clarke from a Banach space E to a Banach manifold M modelled on E.

Let D be a nonempty subset of E and let $u \in D$. Set

$$T_D(u) = \{v \in E;\ \forall h_n \downarrow 0,\ \forall u_n \to u\ (u_n \in D),\ \exists v_n \to v,\ \exists\ u_n + h_n v_n \in D\}. \tag{6.18}$$

This is the tangent cone to D at u in the sense of Clarke. The cone given by (1.6) with E in place of M becomes

$$\tilde{T}_u(D) = \{v \in E;\ \lim_{h\downarrow 0} h^{-1}\, d(u + hv;D) = 0\}. \tag{6.19}$$

Obviously, $T_D(u) \subset \tilde{T}_u(D)$. In general this is a strict inclusion. In view of (2.6) in Chapter 3 it is easily seen that the convexity of D implies the convexity of $\tilde{T}_u(D)$ (see §7). A surprising fact is that Clarke's cone $T_D(u)$ is convex even in the case in which D is not convex. Let us prove this well-known remarkable fact:

<u>Proposition 6.1</u> *Let* D *be a nonempty subset of the Banach space* E. *The cone* $T_D(u)$ *of Clarke is convex.*

<u>Proof</u> Let $v_1, v_2 \in T_D(u)$, $u_n \to u$ (with $u_n \in D$) and $h_n \downarrow 0$. There is $v_n^1 \to v_1$ such that $u_n + h_n v_n^1 \in D$. Furthermore, since $v_2 \in T_D(u)$, there is $v_n^2 \to v_2$ such that $(u_n + h_n v_n^1) + h_n v_n^2 \in D$. If we set $\bar{v}_n = v_n^1 + v_n^2$ then $\bar{v}_n \to v_1 + v_2$ and $u_n + h_n \bar{v}_n$ D (n=1,...). This means that $v_1 + v_2 \in T_D(u)$, q.e.d.

Now let S be a nonempty subset of the manifold M. In an equivalent form. Proposition 1.1 asserts that the tangent cone $\tilde{T}_x(S)$ to S at x (see (1.6)), that is,

$$\tilde{T}_x(S) = (D(\phi)_x)^{-1}\tilde{T}_{\phi(x)}(\phi(U \cap S)) \tag{6.20}$$

where (U,ϕ) is a chart of M at x, is independent of the chart $\phi: U \to \phi(U) \subset E$. Similarly one proves:

<u>Proposition 6.2</u> *The tangent cone*

$$T_S(x) = (D(\phi)_x)^{-1}\ T_{\phi(U\cap S)}(\phi(x)) \tag{6.21}$$

to S *at* x *is independent of the chart* (U,ϕ) *of* M *at* $x \in S$.

<u>Proof</u> Let $\psi: V \to \psi(V) \subset E$ be another chart of M at x and let $y \in T_S(x)$, given by (6.12). This means that $D(\phi)_x y = y_\phi \in T_{\phi(U\cap S)}(\phi(x))$. The only fact we have to prove is that $y_\psi \in T_{\psi(V\cap S)}(\psi(x))$ defined by (6.18). Consequently, let $z_n \to \psi(x)$ and $h_n \downarrow 0$ as $n \to \infty$. Clearly, we may assume that $z_n \in \psi(U \cap V \cap S)$. Set $x_n = \phi\psi^{-1}(z_n) \in \phi(U \cap V \cap S)$. Since $x_n \to \phi(x)$, by hypothesis there exists a sequence $y_n^\phi \to y_\phi$ such that $x_n + h_n^\phi y_n \in \phi(U \cap V \cap S)$. In other words, $\psi\phi^{-1}(x_n + h_n y_n^\phi) \in \psi(U \cap V \cap S)$. On the other hand, by the definition of

Fréchet derivative, we have

$$\psi\phi^{-1}(x_n + h_n y_n^{\phi}) = \psi\phi^{-1}(x_n) + D(\psi\phi^{-1})x_n(h_n y_n^{\phi}) + h_n \|y_n^{\phi}\| r_n$$

$$= z_n + h_n y_n^{\psi} \in \psi(U \cap V \cap S)$$

where $r_n \to 0$ and $y_n \to D(\psi\phi^{-1})_{\phi(x)} y_{\phi} = y_{\psi}$. This completes the proof.

The normal cone $N_D(u)$ to D at $u \in D$ is defined as the dual of $T_D(u)$, that is

$$N_D(u) = \{v^* \in E^*; \langle v^*,v\rangle \leq 0, \quad \forall v \in T_D(u)\}. \tag{6.22}$$

Accordingly, the normal cone $N_S(x)$ to S at $x \in S$ is the dual cone of $T_S(x)$ (in $T_x^*(M)$), hence

$$N_S(x) = \{z^* \in T_x^*(M); \langle z^*,y\rangle \leq 0, \forall y \in T_S(x)\}. \tag{6.23}$$

On the basis of (6.20) it follows that

$$N_S(x) = (D(\phi)_x)^* N_{\phi(U\cap S)}(\phi(x)). \tag{6.24}$$

We now proceed to the definition of generalized gradients of a locally Lipschitz function $f:M \to R$. Recall that f is said to be locally Lipschitz if for every point $x \in M$ there exists a chart $\phi:U \to E$ of M at x, a neighbourhood $W \subset \phi(U)$ of $\phi(x)$ and a positive number $K^{\phi} = K$ such that

$$|f\circ\phi^{-1}(w) - f\circ\phi^{-1}(z)| \leq K\|w - z\|, \quad \forall w,z \in W. \tag{6.25}$$

This property of f does not depend on the chart (U,ϕ). The *generalized directional derivative* $f^0(x;v)$ of f at x in the v direction ($v \in T_x(M)$) is defined by

$$f^0(x;v) = (f\circ\phi^{-1})^0(\phi(x);v_{\phi}) = \lim_{y\to\phi(x);}\ \sup_{h\downarrow 0} (f\circ\phi^{-1}(y+hv_{\phi})-f\circ\phi^{-1}(y))/h \tag{6.26}$$

where $v_{\phi} = D(\phi)_x v$.

<u>Proposition 6.3</u> *The number* $f^0(x;v)$ *is independent of* ϕ.

Proof Let (V,ψ) be another chart of M at x. Set $I_\psi = (f\circ\psi^{-1})^0(\psi(x);v_\psi)$. It suffices to show that $I_\phi \leq I_\psi$. Clearly, (6.26) imply the existence of two sequences $h_n \downarrow 0$ and $y_n \to \phi(x)$ with the property that $(f\circ\phi^{-1}(y_n+h_n v_\phi) - f\circ\phi^{-1}(y_n))/h_n$ converges to I_ϕ. By a standard device we get

$$f\circ\phi^{-1}(y_n + h_n v_\phi) = f\circ\psi^{-1}(\psi\phi^{-1}(y_n + h_n v_\phi))$$

$$= f\circ\phi^{-1}(\psi\phi^{-1}(y_n) + h_n D(\psi\phi^{-1})y_n v_\phi$$

$$+ h_n \|v_\phi\| r_n), \; n = 1,2,\ldots, \; r_n \to 0.$$

Using the Lipschitz property, the notation $z_n = \psi\phi^{-1}(y_n)$ and the formula $v_\psi = D(\psi\phi^{-1})_{\phi(x)}v_\phi$, we obtain

$$f\circ\phi^{-1}(y_n + h_n v_\phi) - f\circ\phi^{-1})_{y_n} \leq Kh_n \|D(\psi\phi^{-1})_{y_n} v_\phi - D(\psi\phi^{-1})_{\phi(x)}v_\phi + \|v_\phi\| r_n\|$$

$$+ (f\circ\psi^{-1}(z_n + h_n v_\psi) - f\circ\psi^{-1}(z_n)).$$

Dividing by h_n and passing to the limit for $n \to \infty$, we get exactly the desired inequality $I_\phi \leq I_\psi$, which completes the proof.

We are now prepared to give:

Definition 6.1 *Let* M *be a* C^k*-manifold, with* $k \geq 1$, *and let* $f:M \to R$ *be a locally Lipschitz function. The generalized gradient* $\partial f(x)$ *of* f *at* $x \in M$ *is the following subset of the cotangent space* $T_x^*(M)$:

$$\partial f(x) = \{y^* \in T_x^*(M);\ f^0(x;v) \geq \langle y^*,v\rangle, \ \forall v \in T_x(M)\} \tag{6.27}$$

where $\langle y^*,v\rangle$ *is the value of* y^* *at* v.

Using a chart $\phi:U \to E$ of M at x, $\partial f(x)$ can also be expressed as

$$\partial f(x) = (D(\phi)_x)^* \, \partial(f\circ\phi^{-1}). \qquad (6.27)'$$

Here the mapping $(D(\phi)_x)^* : E^* \to T_x^*(M)$ is defined by

$$\langle (D(\phi)_x)^*(z^*),y\rangle = \langle D(\phi)_x y,z^*\rangle, \ \forall y \in T_x(M), \ z^* \in E^*. \tag{6.28}$$

Let us consider the problem

minimize $f:M \to R$ subject to: $x \in S \subset M$, $h_j(x) = 0$

$$(j \in J),\ g_i(x) \leqslant 0\ (i \in I) \tag{6.29}$$

where I and J are finite index sets and the functions h_j and g_i are locally Lipschitz from M into R.

The Lagrange multipliers rule in terms of generalized gradients on Banach manifolds has the same form as in the case of a Banach space (i.e., the form given by Clarke [4]), namely:

Theorem 6.1 *Let* M *be a* C^k*-manifold,* $k \geqslant 1$, S *a nonempty subset of* M *and let* $f:M \to R$ *be a locally Lipschitz function. If* $x \in S$ *is a solution to the problem* (6.29), *then there exist scalars* $\lambda \geqslant 0$, $s_j (j \in J)$ *and* $r_i \geqslant 0$ $(i \in I)$ *not all zero with the properties:* $r_i g_i(x) = 0$ *and*

$$\lambda \partial f(x) + \sum_{j \in J} s_j\ \partial h_j(x) + \sum_{i \in I} r_i\ \partial g_i(x) \in -N_S(x). \tag{6.30}$$

Proof Let $\phi:U \to E$ be a chart of M at x. It is clear that the functions $f \circ \phi^{-1}$, $h_j \circ \phi^{-1}$ and $g_i \circ \phi^{-1}$ from $\phi(U)$ into R have similar properties to f, h_j and g_i respectively (with $\phi(x)$ in place of x). Consequently, by Clarke's theorem cited above, there exist $\lambda \geqslant 0$, $s_j (j \in J)$, $r_i \geqslant 0$ $(i \in I)$ not all zero, such that $r_i g_i \circ \phi^{-1}(\phi(x)) = 0$ and

$$\begin{aligned} \lambda \partial (f \circ \phi^{-1})(\phi(x)) &+ \sum_{j \in J} s_j\ \partial h_j \circ \phi^{-1}(\phi(x)) \\ &+ \sum_{i \in I} r_i\ \partial g \circ \phi^{-1}(\phi(x)) \in -N_{\phi(U \cap S)}(\phi(x)). \end{aligned} \tag{6.31}$$

By applying $(D(\phi)_x)^*$ to (6.31) and taking into account (6.24) and (6.27)', one obtains (6.30).

Remark 6.1 In both (6.30) and (6.31) the following convention is adopted: if A, B, C are subsets and s_1, s_2 are scalars, then by $s_1 A + s_2 B \in C$ we mean the fact that there exist $a \in A$ and $b \in B$ such that $s_1 a + s_2 b \in C$.

Remark 6.2 Let us assume that $f:S \subset M \to R$ is defined only on S and that it is locally Lipschitz on S. Then it is clear that in (6.26) we have to replace " $\lim\limits_{y\to\phi(x)}$ ", by " $\lim\limits_{\substack{y\to\phi(x)\\ y+hv_\phi\in\phi(U\cap S)}}$ ". Accordingly, the generalized directional derivative $f^0(x;v)$ is defined in the direction $v \in T_x(M)$ which is quasi-tangent to S at x. Indeed, in this case (i.e., $v \in \tilde{T}_x(S)$), there exists $r = r(h)$ such that (see (1.4) and (1.6))

$$\phi(x) + hv_\phi + hr(h) \in \phi(U \cap S), \quad \forall h > 0. \tag{6.32}$$

Consequently, with $y = y(h) = \phi(x) + hr(h)$, the condition $y + hv_\phi \in \phi(U \cap S)$ is fulfilled.

§7. Notes and Remarks

The results of this chapter are based on Motreanu and the author. For more information in connection with §1, the interested reader should consult Federer [1] and Roger [1].

Note that for proving Formulas (4.5) and (4.7) we have used the transversality of $g:M \to N$ over the submanifold W of N. In Chapter 3, for similar results (i.e., (2.34), (2.35)) we have used Brower's fixed point theorem.

To prove the results of Sections 4 and 5, the crucial remark consists of the fact that differentiability preserves transversality (i.e. Lemma 4.2). Actually, a careful examination of the proof of Lemma 4.2, leads to the following precise result:

Lemma 7.1 *A* C^2*-map* $g:M \to N$ *is transversal over the submanifold* W *of* N, *if and only if* D(g) *is transversal over* T(W).

For Theorem 4.1 we refer to Lang [1]. We do not have a reference for Lemma 4.2. It is easy to see that a Banach manifold M is a very natural framework in which to treat flow-invariance problems. Moreover, in this case we can use some powerful tools of the manifolds. We think that our comparison here, between the techniques of Chapters 3 and 4, is useful. It is also interesting to point out that the flow-invariance of a subset S with respect to a second-order equation is an important problem of dynamics, which leads to the necessity of considering the transversality of both g (over W) and D(g) (over TW).

The dissipativity condition cannot be extended to Banach manifolds, since this notion is not independent of charts. Consequently, some results in

Chapter 2 (e.g., Theorem 2.1) cannot be extended to manifolds.

We have also to note that the condition $D(\pi)_y z = y$, from Lemma 3.1, is essential. Without it, the property (3.5) is not independent of the chart (U,ϕ).

The purpose of Section 6 is to show that (as in the case of a Banach space) the notion of quasi-tangency has also a unifying effect in optimization. In connection with Clarke's cone $T_D(u)$ (see (6.18)) we have to add that it is also closed. Finally, in connection with $\tilde{T}_u(D)$ given by (6.19), we want to recall:

Proposition 7.1 (1) *The "contingent cone"* $\tilde{T}_u(D)$ *of* D *at* u *is closed.* (2) *If* D *is convex then* $\tilde{T}_u(D)$ *is also convex.*

Proof If $v \in \tilde{T}_u(D)$, then for every $\lambda > 0$, $\lambda v \in \tilde{T}_u(D)$, that is, $\tilde{T}_u(D)$ is certainly a cone. (1) Now, let $v_p \in \tilde{T}_u(D)$ be such that $\|v_p - y\| \to 0$ as $p \to \infty$. We have to show that $y \in \tilde{T}_u(D)$. This follows from the inequality

$$d(u+hy;D) = |d(u+hy;D) - d(u+hv_p;D)| + d(u+hv_p;D)$$

$$\leq h\,\|y-v_p\| + d(u+hv_p;D)$$

which implies

$$\limsup_{h\downarrow 0} h^{-1}d(u+hy;D) \leq \|v_p-y\|\,,\ p = 1,2,\ldots\,.$$

(2) Let $v_1, v_2 \in \tilde{T}_u(D)$. This is equivalent to the existence of $r_1(h), r_2(h) \to 0$ as $h \downarrow 0$ such that $u + h(v_q + r_q(h)) \in D$, $\forall h > 0$, $q = 1,2$. Then the convexity of D yields

$$u + h[(\lambda v_1 + (1-\lambda)v_2) + (\lambda r_1(h) + (1-\lambda)r_2(h))] \in D,\ \forall\lambda \in [0,1],$$

that is, $\lambda v_1 + (1-\lambda)v_2 \in T_u(D)$.

5 Perturbed differential equations on closed subsets

§1. Differential inclusions with linear compact semigroup generators. Pazy's local existence theorem

Let X be a Banach space of norm $\|\cdot\|$ and let X* be its dual.

Definition 1.1 *A one-parameter family of bounded linear operators* $S(t):X \to X$, $t \geq 0$, *is a semigroup of class* C_0 *(or simply a* C_0 *semigroup) if:*

(i) $S(0) = I$ *(the identity operator on* X).

(ii) $S(t + s) = S(t)S(s)$, $\forall t,s \geq 0$ *(the semigroup property).*

(iii) $\lim_{t\downarrow 0} S(t)x = x$, $\forall x \in X$ *(strong continuity).*

It is well known that if S(t) is a C_0 semigroup on X, there are constants $C \geq 1$ and $\omega \geq 0$ such that

$$\|S(t)\| \leq C\, e^{\omega t}, \text{ for } 0 \leq t \to \infty . \tag{1.1}$$

The linear operator A defined by

$$Ax = \lim_{t\downarrow 0} \frac{S(t)x - x}{t}, \; x \in D(A) \tag{1.2}$$

$$D(A) = \{x \in X;\ \lim_{t\downarrow 0} \frac{S(t)x - x}{t} \text{ exists}\} \tag{1.3}$$

is said to be the *infinitesimal generator* of the semigroup S(t). If C = 1 and $\omega = 0$, then S(t) is said to be a C_0 *semigroup of contractions*. According to the Hille-Yosida theorem:

A linear (unbounded) operator A *is the infinitesimal generator of a* C_0 *semigroup of contractions* S(t), $t \geq 0$, *if and only if* $\overline{D(A)} = X$, A *is closed, the resolvent set* $\rho(A) \supset R_+$ *and*

$$\|(\lambda I - A)^{-1}\| \leq \lambda^{-1}, \quad \forall \lambda > 0. \tag{1.4}$$

In this section we are concerned with the differential equation

$$u' \in Au + F(t,u),\ t \in]a,b[,\ u(t_0) = x_0 \in D(t_0) \tag{1.5}$$

where A is the generator of a C_0 semigroup S(t), with S(t) compact S(t) for $t > 0$ (i.e., for $t > 0$, S(t) maps bounded subsets into relatively compact subsets). Our hypotheses here allow the possibility that F is multivalued. For this reason, Equation (1.5) is also called a "differential inclusion". Moreover, the domain

$$D(F(t,\cdot)) = D(t) \subset X,\ t \in]a,b[,\ -\infty \leqslant a < b \leqslant +\infty \tag{1.6}$$

of the (possible multivalued) operator $x \to F(t,x)$ may be time-dependent.

For the statement of the results let us recall:

Definition 1.2 *The operator* $(t,x) \to F(t,x)$ *(possible multivalued) is said to be locally bounded if for each* $t \in]a,b[$ *and* $x \in D(t)$ *there are constants* $T, M, r > 0$ *such that:* $a < t-T$, $t + T < b$ *and*

$$||f(s,y)|| \leqslant M,\ \forall s \in [t-t,t+T],\ y \in B(x,r) \cap D(s), f(s,y) \in F(s,y).$$

F *is said to be demiclosed if the hypotheses:* $y_n \in F(t_n,x_n)$ *with* $x_n \in D(t_n)$, $t_n \to t$, $x_n \to x$ *and* $y_n \to y$, *imply* $x \in D(t)$ *and* $y \in F(t,x)$.

Definition 1.3 *Let* J *be a subinterval of* $[t_0,b[$ *with* $t_0 \in J$. *A continuous function* $u:J \to X$ *is said to be a mild solution to the initial value problem* (1.5) *on* J *if* $u(t) \in D(t)$ *for all* $t \in J$ *and if there exists a strongly measurable selection* f *of* F *on* J *(i.e.,* $f(s) \in F(s,u(s))$ *a.e. on* J *with* f *measurable from* J *into* X*) such that*

$$u(t) = S(t-t_0)x_0 + \int_{t_0}^{t} S(t-s)f(s)ds,\quad \forall t \in J. \tag{1.7}$$

For convenience of future reference we record the following conditions:

(C1) A: $D(A) \subset X \to X$ is the infinitesimal generator of the C_0 semigroup S(t). For $t > 0$, S(t) is compact.

(C2) For each $t_0 \in]a,b[$ and $x \in D(t_0)$ there exist $r > 0$ and $\bar{T} \in]t_0,b[$ such that $B(x,r) \cap D(t)$ is nonempty for all $t \in [t_0,\bar{T}]$ and the mapping $t \to B(x,r) \cap D(t)$ is closed on $[t_0,\bar{T}]$.

(C3) The mapping $(t,x) \to F(t,x)$ is nonempty, convex and closed valued, demiclosed and locally bounded.

(C4) $\lim_{h\downarrow 0} h^{-1} d(S(h)x + hf(t,x);D(t+h)) = 0$, for all $t \in]a,b[$, $x \in D(t)$ and $f(t,x) \in F(t,x)$.

Using the nonexpansivity of the distance function (Ch. 2, (2.3)'), it is easy to check that (C4) is equivalent to

(C4)' $\lim_{h\downarrow 0} h^{-1} d(S(h)x + \int_t^{t+h} S(t+h-s)f(t,x)ds;D(t+h)) = 0$,

for all $t \in]a,b[$, $x \in D(t)$ and $f(t,x) \in F(t,x)$.

Remark 1.1 In the case $D(t) = D$ independent of t, Condition (C2) means that D is locally closed and (C4) becomes

(C4)" $\lim_{h\downarrow 0} h^{-1} d(S(h)x + hf(t,x);D) = 0$,

for all $t \in]a,b[$, $x \in D$ and $f(t,x) \in F(t,x)$.

Clearly, if D is an open subset Condition (C4)" is trivially satisfied, because in this case for each $x \in D$ it follows that $S(h)x + hf(t,x) \in D$ for all sufficiently small $h > 0$ and $f(t,x) \in F(t,x)$ (with F locally bounded).

The characterization of the operators satisfying Condition (C1) is given by the following theorem of Pazy [1]:

Theorem 1.1 (Pazy). *The operator* $A:D(A) \subset X \to X$ *is the infinitesimal generator of the compact* C_0 *semigroup* $S(t)$, *if and only if* $S(t)$ *is continuous in the uniform operator topology for* $t > 0$ *and the resolvent* $(\lambda I-A)^{-1}$ *is compact for* $\lambda \in \rho(A)$.

Remark 1.2 Let X be a Banach space of infinite dimension. In this case, I is not compact. Consequently (C1) implies that A is different from the null operator. Moreover, on the basis of Pazy's theorem and of $(I-A)(I-A)^{-1} = I$, we see that (C1) also excludes the continuity of A. Other comments are given in §5.

We now are prepared to state the main result of this section:

Theorem 1.2 *Let X be a reflexive space and let us assume that Conditions*

(C1)-(C4) *are fulfilled. Then for each* $t_0 \in]a,b[$ *and* $x_0 \in D(t_0)$ *there exists at least a mild solution* u *to the initial value problem* (1.5) *on* $]t_0,t_0+T]$ (*with* $T = T(t_0,x_0) > 0$ *and* $t_0 + T < b$).

For the proof of this theorem the following result due to Kato is needed:

Lemma 1.1 *Let* X *be reflexive and let* u_n *be a sequence of* $L^p(0,T;X)$, $p > 1$, *such that* $u_n(t)$ *is bounded a.e. on* $[0,T]$ (*i.e., for almost all* $t \in [0,T]$ *there exists* $K(t) > 0$ *with* $\|u_n(t)\| \leq K(t)$, $n = 1,2,\ldots$). *If* $u_n \longrightarrow u$ *(weakly) in* $L^p(0,T;X)$, *then* $u(t) \in \text{clco } Y(t)$ *a.e. on* $[0,T]$, *where* $Y(t)$ *is the set of all weak cluster points of* $u_n(t)$, *while* clco *denotes the closure of the convex hull.*

Proof Since $u_n \longrightarrow u$, it follows that $u \in \text{clco } \{u_n\}_1^\infty$. Accordingly, there exists a sequence v_n of convex combinations of $u_1,u_2,\ldots$ with v_n strongly convergent to u in $L^p(0,T;X)$. We may assume (relabelling if necessary) that $v_n(t) \to u(t)$ (strongly in X) as n a.e. on $[0,T]$. Given $t \in [0,T]$ let E be an arbitrary open half-space of X containing $Y(t)$. Then $u_n(t) \in E$ for all large enough n, which implies that $v_n(t) \in E$ for sufficiently large n. Therefore $\lim_{n\to\infty} v_n(t) = u(t) \in E$. The conclusion of this lemma follows from the fact that clco $Y(t)$ is the intersection of all closed half-spaces E which contain it.

Another preliminary result is:

Proposition 1.1 *Suppose that the hypotheses of Theorem* 1.2 *hold. Let* $t_0 \in]a,b[$ *and* $x_0 \in D(t_0)$. *Choose* $r > 0$ *and* $\bar{T} = t_0 + T$ *satisfying* (C2) *as well as*

$$\|f(t,x)\| \leq M, \ \forall t \in [t_0,\bar{T}], \ x \in D(t) \cap B(x_0,r), f(t,x) \in F(t,x). \quad (1.8)$$

Moreover, assume that T *is sufficiently small such that*

$$\max_{0\leq t\leq T} \|S(t)x - x\| + T(M+1)N \leq r, \text{ where } N = C\, e^{\omega T}. \quad (1.9)$$

Then there is an $\frac{1}{n}$*-approximate solution* u_n *to* (1.1) *on* $[t_0,\bar{T}]$ *in the following sense:*

For each positive integer n, *there is an infinite partition* $\{t_i^n\}_{i=0}$ *of* $[t_0,\bar{T}]$ *with the properties:*

(a1) $t_0^n = t_0$, $t_{i+1}^n - t_i^n \equiv d_i^n \in]0,1/n]$, $\lim_{i\to\infty} t_i^n = \bar{T}$

(a2) $u_n(t_0) = x_0$, $u_n(t_i^n) \equiv x_i^n \in B(x_0,r) \cap D(t_i^n)$

(a3) $u_n(t) = S(t-t_i^n)x_i + \int_{t_i^n}^{t} S(t-s)f(t_i^n,x_i^n)ds + (t-t_i^n)p_i^n \in B(x_0,r)$

for $t \in [t_i^n, t_{i+1}^n]$ *with* $\|p_i^n\| \leq \frac{1}{n}$ *for* $i = 0,1,\ldots$ *where* $f(t_i^n,x_i^n)$ *is an element of* $F(t_i^n,x_i^n)$.

Proof The construction of u_n and $\{t_i^n\}$ is by induction on i. Therefore set $t_0^n = t_0$, $u_n(t_0) = x_0$ and assume that u_n is constructed on $[t_0^n,t_i^n]$. To simplify notation, supress n as a superscript for t_i, d_i, x_i and p_i. If $t_i = \bar{T}$, set $t_{i+1} = t_i$, and if $t_i < \bar{T}$, define

$$\delta_i = \sup \{h \in]0,1/n];\ t_i + h \leq \bar{T},\ \text{there is } f(t_i,x_i) \in F(t_i,x_i),$$
$$d(S(h)x_i + \int_{t_i}^{t_i+h} S(t_i+h-s)f(t_i,x_i)ds;D(t_i+h)) \leq h/2n\}. \tag{1.10}$$

Consequently, there exist $d_i \in]2^{-1}\delta_i,\delta_i]$ (with $0 < d_i \leq 1/n$, $t_i + d_i \leq T$) and $f(t_i,x_i) \in F(t_i,x_i)$ such that

$$d(S(d_i)x_i + \int_{t_i}^{t_i+d_i} S(t_i+d_i-s)f(t_i,x_i)ds;D(t_i + d_i)) \leq \frac{d_i}{2n}\,. \tag{1.10)'$$

Set $t_{i+1} = t_i + d_i$. By (1.10)' we see that there is an element $x_{i+1} \in D(t_{i+1})$ with the property

$$\|S(d_i)x_i + \int_{t_i}^{t_{i+1}} S(t_{i+1} - s)f(t_i,x_i)ds - x_{i+1}\| \leq \frac{d_i}{n}. \tag{1.11}$$

It follows from (1.11) that x_{i+1} can be written under the form (see also Ch. 2, (2.8))

$$x_{i+1} = S(t_{i+1} - t_i)x_i + \int_{t_i}^{t_{i+1}} S(t_{i+1} - s)f(t_i,x_i)ds + (t_{i+1}-t_i)p_i \tag{1.12}$$

with $\|p_i\| \leq 1/n$. We now define u_n on $[t_i,t_{i+1}]$ as indicated by (a3). By induction on i (which is left to the reader) one verifies that

$$u_n(t) = S(t-t_0)x_0 + \sum_{j=0}^{i-1} \int_{t_j}^{t_{j+1}} S(t-s)f(t_j,x_j)ds + \int_{t_i}^{t} S(t-s)f(t_i,x_i)ds$$

$$+ \sum_{j=0}^{i-1} (t_{j+1}-t_j)S(t-t_{j+1})p_j + (t-t_i)p_i, \text{ for}$$

$$t \in [t_i,t_{i+1}]. \tag{1.13}$$

Let us define the step function:

$$a_n(s) = t_i \text{ for } s \in [t_i,t_{i+1}], \; a_n(\bar{T}) = \bar{T}. \tag{1.14}$$

Then one has $u_n(a_n(s)) = u_n(t_i^n) = x_i^n$ for $s \in [t_i,t_{i+1}[$. Set

$$g_n(t) = \sum_{j=0}^{i-1} (t_{j+1} - t_j)S(t-t_{j+1})p_j + (t-t_i)p_i, \text{ for } t \in [t_i,t_{i+1}] \tag{1.15}$$

(so $\|g_n(t)\| \leqslant N(t-t_0)/n$). Finally set

$$f_n(s) = f(a_n(s), u_n(a_n(s))) = f(t_i,x_i), \text{ for } s \in [t_i,t_{i+1}[. \tag{1.16}$$

Then u_n can be written in the form

$$u_n(t) = S(t-t_0)x_0 + \int_{t_0}^{t} S(t-s)f_n(s)ds + g_n(t), \tag{1.17}$$

for all t in the domain of u_n. Actually (1.17) holds for all $t \in [t_0,\bar{T}]$, since $\lim_{i\to\infty} t_i = \bar{T}$ (we shall prove this in the sequel). For a moment (by induction hypothesis) we have $u_n(t) \in B(x_0,r)$ for all $t \in [t_0,t_i]$. Then by (1.13) (or (1.17)) and (1.9),

$$\|u_n(t)-x_0\| \leqslant \|S(t-t_0)x_0-x_0\| + TMN + TN/n \leqslant r, \; \forall t \in [t_i,t_{i+1}].$$

Thus Properties (a2) and (a3) are verified.

Since the proof of the last property in (a1) is quite difficult, we break it into two steps, as follows: *The convergence of* $\{x_i\}_{i=0}^{\infty}$. Since $t_i \leqslant T$, $\lim_{i\to\infty} t_i = t \leqslant T$ exists. We now prove that $\lim_{i\to\infty} x_i = x'$ exists too. Precisely, we show that $\{x_i\}$ is a Cauchy sequence. Indeed, let $j > i$. Using

(1.13) one easily obtains

$$\|x_j - x_i\| \leq N\,\|S(t_j-t_i)x_o - x_o\| + \sum_{m=0}^{i-1} \|S(t_j-t_i)-I)f(t_m,x_m)\|\,(t_{m+1}-t_m)N$$

$$+\, N \sum_{m=0}^{i-1} \|S(t_j-t_i)p_m - p_m\|\,(t_{m+1}-t_m) + (M+1)(t_j-t_i)N. \tag{1.18}$$

Let $\varepsilon > 0$ be given. There is a positive integer $k = k_\varepsilon$ such that (with $N = 1$, for simplicity)

$$t_j - t_i < \varepsilon'/6(M+1),\ \|S(t_j-t_i)x_o - x_o\| \leq \varepsilon'/6, \text{ for all } j \geq i \geq k \tag{1.19}$$

where $\varepsilon' = \varepsilon/2$. Choose $\bar{k} > k$, with the property

$$\begin{aligned} &\|S(t_j-t_i)p_m - p_m\| \leq \varepsilon'/3\bar{T} \\ &\|S(t_j-t_i) - I)f(t_m,x_m)\| \leq \varepsilon'/3\bar{T} \end{aligned} \tag{1.20}$$

for $m = 0,1,\ldots,k-1$ and $j \geq i \geq \bar{k}$. Denote by I_q the q-th term ($q = 1,2,3,4$) of the right-hand side of Inequality (1.18). Then by (1.19) and (1.20) we have

$$I_2 \leq \varepsilon' t_k/3\bar{T} + 2M(t_i - t_k) \leq 2\,\varepsilon'/3,\quad j \geq i \geq \bar{k}$$

$$I_3 \leq \varepsilon' t_k/3\bar{T} + 2(t_i - t_k)/n \leq \varepsilon'/3 + \varepsilon'/3n(M+1) \leq \varepsilon'/2$$

for $n > 3/(M+1)$, $j \geq i \geq \bar{k}$. It is clear that (1.18) yields

$$\|x_j - x_i\| \leq \varepsilon'/6 + 2\,\varepsilon'/3 + \varepsilon'/2 + \varepsilon'/6 < \varepsilon$$

for all $j \geq i \geq \bar{k}$ and therefore $\lim_{i\to\infty} x_i = x^*$ exists. Combining (C2), (C3) and (a2) it follows that $x^* \in B(x_o,r) \cap D(t^*)$. *The convergence of* $\{t_i\}$ *to* $\bar{T}$. The proof of the fact that $\lim_{i\to\infty} t_i = \bar{T}$ is by contradiction. Therefore, let us assume that $\lim_{i\to\infty} t_i = t^* < \bar{T}$. According to (1.8) we have $\|f(t_i,x_i)\| \leq M$ for all $i = 1,2,\ldots$. Since X is reflexive, we may suppose (relabelling if necessary) that $f(t_i,x_i)$ is weakly convergent to an element f^*. By the construction of $f(t_i,x_i)$ we have $f(t_i,x_i) \in F(t_i,x_i)$, which yields $f^* \in F(t^*,x^*)$

(because F is demiclosed). Then, in view of (C4), we can choose $h^* \in]0,1/n[$ with the properties: $t^* + h^* < \bar{T}$ and

$$d(S(h^*)x^* + \int_{t^*}^{t^*+h^*} S(t^* + h^*-s)f^*ds; D(t^* + h^*)) \leq \frac{h^*}{4n}. \qquad (1.21)$$

Let us observe that $\lim_{i\to\infty} d_i = \lim_{i\to\infty} (t_{i+1} - t_i) = 0$. Hence $2^{-1}\delta_i < d_i$ implies $\delta_i < h^*$ for all large enough i (say, $i \geq i_0$). Set $h_i = t^*-t_i + h^*$. Then $t_i + h_i = t^* + h^*$, $h_i > \delta_i$ and $h_i \uparrow h^*$ as i goes to infinity. It is now clear that, according to Definition (1.10) of δ_i, it follows that

$$d(S(h_i)x_i + \int_{t_i}^{t_i+h_i} S(t_i+h_i-s)f(t_i,x_i)ds; D(t_i+h_i)) > \frac{h_i}{2n} \qquad (1.22)$$

for all $i \geq i_0$. We now consider the sequences

$$v_i = S(h_i)x_i + \int_{t_i}^{t_i+h_i} S(t_i+h_i-s)f(t_i,x_i)ds \qquad (1.23)$$

and, for $0 < \varepsilon < h^*$ ($h^* < h_i$, $i \geq i_0$),

$$v_i(\varepsilon) = S(h_i)x_i + \int_{t_i}^{t_i+h_i-\varepsilon} S(t_i+h_i-s)f(t_i,x_i)ds, \; i \geq i_0. \qquad (1.24)$$

Since 0 $t_i+h_i-s-\varepsilon < h_i - \varepsilon$ for all $i \geq i_0$, we have

$$v_i(\varepsilon) = S(h_i)x_i + S(\varepsilon) \int_{t_i}^{t_i+h_i-\varepsilon} S(t_i+h_i-s-\varepsilon)f(t_i,x_i)ds, \; i \geq i_0. \; (1.25)$$

On the basis of the compactness of $S(\varepsilon)$ as well as the fact that $\lim_{i\to\infty} S(h_i)x_i = S(h^*)x^*$, (1.25) implies the precompactness of $\{v_i(\varepsilon)\}_{i=0}^{\infty}$. On the other hand, $\|v_i-v_i(\varepsilon)\| \leq M\varepsilon$ for $i \geq i_0$, which shows that $\{v_i\}$ is also precompact (or relatively compact) in X. Set

$$v^* = S(h^*)x^* + \int_{t^*}^{t^*+h^*} S(t^* + h^* - s)f^* \, ds.$$

Recalling that $f(t_i,x_i) \rightharpoonup f^*$ as $i \to \infty$, we see that $v_i \rightharpoonup v^*$ as $i \to \infty$. Since we may assume that $\{v_i\}$ is strongly convergent, then (by the uniqueness of the weak limit) it follows that $v_i \to v^*$ as $n \to \infty$. Finally, $t_i+h_i = t^*+h^*$,

$h_i \downarrow h^*$, (1.21) and (1.22) give us a contradiction (letting $i \to \infty$ in (1.22)) which concludes the proof.

Proof of Theorem 1.2 We shall prove that the sequence u_n constructed via Proposition 1.1 has a strongly convergent subsequence in the norm of $C([t_0,\bar{T}];X)$ (to a mild solution of (1.5)). To this end we shall apply the infinite dimensional version of Ascoli-Arzela's theorem. First of all, by (a3) we have $u_n(t) \in B(x_0,r)$ for all $t \in [t_0,\bar{T}]$. In view of (1.17) it suffices to show that

$$y_n(t) = \int_{t_0}^{t} S(t-s)f_n(s)ds$$

is equicontinuous on $[t_0,\bar{T}]$. Let us prove this fact with $t_0 = 0$ (for simplicity of writing).

By standard arguments we derive

$$\|y_n(t)-y_n(s)\| \leqslant MN(t-s) + \int_0^s \|S(t-\tau)f_n(\tau)-S(s-\tau)f_n(\tau)\| \, d\tau$$

$$\leqslant MN(t-s) + M\int_0^s \|S(t-\tau)-S(s-\tau)\| \, d\tau, \quad 0 < s < t \leqslant T. \tag{1.26}$$

Let $\varepsilon \in]0,s/2[$. Then we have

$$\int_0^s \|S(t-\tau)-S(s-\tau)\| \, d\tau \leqslant \int_\varepsilon^{s-\varepsilon} \|S(t-\tau)-S(s-\tau)\| \, ds + 4\varepsilon. \tag{1.27}$$

In view of Theorem 1.1 of Pazy, $t \to S(t)$ is continuous (in the uniform operator topology) at each $t > 0$. Consequently,

$$\int_\varepsilon^{s-\varepsilon} \|S(t-\tau)-S(s-\tau)\| \, d\tau = \int_\varepsilon^{s-\varepsilon} \|S(t-s+\tau) - S(\tau)\| \, d\tau$$

$$< (s-2\varepsilon)\varepsilon \quad \text{for } t-s < \delta(\varepsilon). \tag{1.28}$$

It follows from (1.26) - (1.28) that $\{t \to y_n(t)\}_{n=0}^{\infty}$ are equicontinuous on $[0,\bar{T}]$. We now prove that for each $t \in [0,\bar{T}]$ the set $\{y_n(t)\}_{n=0}^{\infty}$ is precompact in X. For $t = 0$, $y_n(0) = 0$ so we have to consider $t > 0$. For $\varepsilon \in]0,t[$, set

$$y_n^\varepsilon(t) = \int_0^{t-\varepsilon} S(t-s)f_n(s)ds = S(\varepsilon)\int_0^{t-\varepsilon} S(t-s-\varepsilon)f_n(s)ds.$$

It is clear that $\{y_n^\varepsilon(t)\}_{n=0}^\infty$ is precompact. This fact and the inequality $\|y_n^\varepsilon(t) - y_n(t)\| \leq MN\varepsilon$ imply the precompactness of $\{y_n(t)\}$, as we wanted. Accordingly, by Ascoli-Arzela's theorem we may assume that u_n is convergent in $C([t_0,\bar{T}];X)$ to a continuous function u.

It remains to prove that u is a mild solution to (1.5). First of all we have to prove that u is D-valued. For this purpose let us consider an arbitrary $s \in [t_0,\bar{T}]$ and an arbitrary positive integer n. There is a positive integer $i_n = i_n(s)$ such that $s \in [t_{i_n}, t_{i_n+1}[$. Thus

$$s - t_{i_n} < t_{i_n+1} - t_{i_n} \leq 1/n,\ a_n(s) = t_{i_n} \text{ and } u_n(a_n(s)) = x_{i_n}.$$

Therefore $|a_n(s) - s| \leq 1/n$ and

$$\|u_n(a_n(s)) - u(s)\| \leq \|x_{i_n} - u(s)\|$$

$$\leq \|u_n(t_{i_n}) - u_n(s)\| + \|u_n(s) - u(s)\|$$

which yields $\lim_{n\to\infty} x_{i_n} = u(s)$. On the other hand, $t_{i_n} \to s$ and $x_{i_n} \in B(x_0,r) \cap D(t_{i_n})$ imply $u(s) \in B(x_0,r) \cap D(s)$.

To complete the proof let us observe that $f_n(s) \in F(a_n(s),u_n(a_n(s)))$ and $\|f_n(s)\| \leq M$ for all $s \in [t_0,\bar{T}]$ and $n = 1,2,\ldots$. Without loss of generality we may assume that $f_n \rightharpoonup f$ (weakly) in $L^2([t_0,\bar{T}];X)$. Denote by Y(s) the set of all cluster points of $\{f_n(s)\}_{n=0}^\infty$. In view of Lemma 1.1, $f(s) \in \text{clco } Y(s)$ a.e. on $[t_0,\bar{T}]$. On the other hand, $Y(s) \subset F(s,u(s))$ (since F is demiclosed). Finally, since $F(s,u(s))$ is closed and convex, it follows that $F(s) \in F(s,u(s))$ a.e. on $[t_0,\bar{T}]$ so f is a strongly measurable selection of $F(\cdot,u(\cdot))$. We may assume that y_n is convergent in $C([t_0,\bar{T}];X)$ (say, to y). In view of (1.17) we have to prove that

$$y(t) = \int_{t_0}^{t} S(t-s)f(s)ds,\ t \in [t_0,\bar{T}]. \tag{1.29}$$

Indeed, let x* be arbitrary in X*. Then

$$\langle y_n(t),x^*\rangle = \int_{t_0}^{t} \langle S(t-s)f_n(s),x^*\rangle ds$$

$$= \int_{t_0}^{t} \langle f_n(s),S^*(t-s)x^*\rangle ds$$

where $S^*(t-s)$ is the adjoint of $S(t-s)$. Letting $n \to \infty$ one obtains (1.29). Clearly, (1.17) is actually

$$u_n(t) = S(t-t_o)x_o + y_n(t) + g_n(t),\ t \in [t_o,\bar{T}]$$

with $\|g_n(t)\| \leqslant NT/n$. Passing to the limit for $n \to \infty$, we get (1.7) with $J = [t_o,\bar{T}]$, which concludes the proof.

In particular, if F is single valued, then (1.5) can be rewritten as

$$u'(t) = Au(t) + F(t,u(t)),\ u(t_o) = x_o \in D(t_o). \tag{1.30}$$

A mild solution to (1.30) on $J \subset [t_o,b]$, $t_o \in J$, is a continuous function u on J which satisfies

$$u(t) = S(t-t_o)x_o + \int_{t_o}^{t} S(t-s)F(s,u(s))ds,\ u(t) \in D(t),\ t \in J. \tag{1.31}$$

In this case Theorem 1.2 reduces to:

Theorem 1.3 *Let* X *be reflexive and let* (C1) *and* (C2) *be fulfilled. Suppose in addition that* F *is demiclosed, locally bounded and satisfies the tangential condition below:*

$$\lim_{h\downarrow 0} h^{-1}\, d(S(h)x + hF(t,x);D(t+h)) = 0,\quad \forall t \in]a,b[,\ x \in D(t). \tag{1.32}$$

Then for each $t_o \in]a,b[$ *and* $x_o \in D(t_o)$ *there exist* $T = T(t_o,x_o) > 0$ *with* $t_o + T < b$, *and a continuous function* u *on* $[t_o,t_o + T]$ *satisfying* (1.31) *with* $J = [t_o,t_o + T]$.

If F is continuous (i.e., $t_n \to t$, $x_n \in D(t_n)$, $x_n \to x$ and $x \in D(t)$ imply $F(t_n,x_n) \to F(t,x)$) then the conclusion of Theorem 1.3 remains valid without the reflexivity assumption upon X. Let us now proceed to state the result in a precise form:

Theorem 1.4 *Let* X *be a general Banach space and* F *a continuous function at every* (t,x) *with* $t \in]a,b[$ *and* $x \in D(t)$. *Suppose that* (C1) *and* (C2) *also holds. Then for each* $t_o \in]a,b[$ *and* $x_o \in D(t_o)$ *there exists a local solution* u *to* (1.31) *on* $[t_o,t_o + T(t_o,x_o)] \subset [t_o,b[$ *if and only if the tangential condition* (1.32) *is fulfilled.*

Proof *The sufficiency of* (1.32). A quick check of the proof of Theorem 1.2 shows that the reflexivity of X has been used to prove that $f(s) \in F(s,u(s))$, only. But we now have $f_n(s) = F(a_n(s),u_n(a_n(s)))$ with $\lim_{n\to\infty} (a_n(s),u_n(a_n(s))) = (s,u(s))$ uniformly with respect to $s \in [t_0,t_0 + T(t_0,x_0)]$. Using this fact, the continuity of F and passing to the limit in (1.17) for $n \to \infty$, we get (1.31).

The necessity of (1.32). Let $t_0 \in]a,b[$, $x_0 \in D(t_0)$ and let u be a continuous function on $[t_0,t_0 + T] = J$ satisfying (1.31). Then for all $h > 0$ with $h < T$, $u(t_0 + h) \in D(t_0 + h)$. Hence

$$
\begin{aligned}
h^{-1} & d(S(h)x_0 + hF(t_0,x_0);D(t_0 + h)) \\
& \leq h^{-1} \|S(h)x_0 + hF(t_0,x_0) - u(t_0 + h)\| \\
& \leq \|F(t_0,x_0) - h^{-1} \int_{t_0}^{t_0+h} S(t_0+h-s)F(s,u(s))ds\| .
\end{aligned}
$$

Letting $h \downarrow 0$, one obtains (1.32) with t_0 and x_0 in place of t and x, respectively. The proof is complete.

In the case $D(t) = D$ independent of t, (C2) means that D is locally closed. Consequently, Theorem 1.4 yields:

Theorem 1.5 *Let* $D \subset X$ *be a locally closed subset of the Banach space* X, $F:]a,b[\times D \to X$ *a continuous function and let* A *be the infinitesimal generator of a* C_0 *semigroup* $S(t)$, $t \geq 0$, *with* $S(t)$ *compact for* $t > 0$. *A necessary and sufficient condition for Equation* (1.30) *to have a local mild solution* $u:[t_0,t_0 + T(t_0,x_0)] \to D$ $(0 < T(t_0,x_0) < b-t_0)$, *for every* $t_0 \in]a,b[$ *and* $x_0 \in D$, *is the following tangential condition:*

$$\lim_{h\downarrow 0} h^{-1} d(S(h)x + hF(t,x);D) = 0, \quad \forall t \in]a,b[,\ x \in D. \tag{1.33}$$

It should be noted that in the case in which $x \in D(A) \cap D$, then (1.33) implies

$$\lim_{h\downarrow 0} h^{-1} d(x + h(Ax + F(t,x));D) = 0. \tag{1.34}$$

This is a consequence of (1.2) along with the nonexpansivity of the distance function.

Remark 1.3 If (1.33) holds with "$h \to 0$" in place of "$h \downarrow 0$", then the solution u to (1.30) exists both to the right and to the left of t_o. Of course, (1.33) guarantees $u(t) \in D$ to the right of t_o, only. Moreover, if D is open, then (1.33) is automatically satisfied and Theorem 1.5 reduces to Pazy's local existence theorem. Finally, if X is finite dimensional and D is open, we can take $A = 0$ (i.e., $S(t) = I$, $\forall t \geq 0$). In this case, in view of (1.31), Theorem 1.5 is just the classical Peano's existence theorem.

Before proceeding to point out the relationship between mild solutions and strong solutions, the following definition is needed:

Definition 1.4 *The function* $u:[t_o,\bar{T}[\to X$, $t_o < \bar{T} < b$, *is a strong (classical) solution to* (1.30) *on* $[t_o,\bar{T}[$ *if* u *is continuously differentiable on* $]t_o,\bar{T}[$, *continuous on* $[t_o,\bar{T}[$, $u(t) \in D(A) \cap D(t)$ *for* $t \in [t_o,\bar{T}[$ *and* (1.30) *is satisfied on* $]t_o,\bar{T}[$.

It is easy to check that a strong solution to (1.30) is also a mild solution. The converse statement need not be true even if $F = 0$. Indeed, if $F = 0$ then the mild solution to (1.30) is just $u_o(t) = S(t-t_o)x_o$ which need not be differentiable when $x_o \bar{\in} D(A)$. However, if, for example, F is continuously differentiable from $[t_o,\bar{T}[\times X$ into X, then the mild solution u to (1.30) is the (unique) strong solution. For such results, a good reference work is the recent book of Pazy [5].

We conclude this section with the statement of Pazy's [3] local existence theorem, as indicated by Remark 1.3. Namely, if D is open (say $D = U$), then Theorem 1.5 reduces to:

Theorem 1.6 (Pazy). *Let* X *be a Banach space and* $U \subset X$ *be open. Let* A *be the infinitesimal generator of a compact semigroup* $S(t)$. *If* $F:]a,b[\times U \to X$ *is continuous, then for every* $t_o \in]a,b[$ *and* $x_o \in U$, *Problem* (1.31) *has a local solution on* $[t_o,t_o + T]$, (*with* $T = T(t_o,x_o) > 0$).

§2. Semilinear differential equations of retarded type. An extension of Hale's theorem

Throughout this section we retain (on A) the notation and hypotheses of the previous section. Let q and T be positive numbers and $-\infty \leq a < b \leq +\infty$. Given a real number t_o, a function $x:[t_o-q,t_o+T] \to X$ and $t \in [t_o,t_o+T]$, define

$x_t:[-q,0] \to X$ by $x_t(\theta) = x(t+\theta)$, for all $\theta \in [-q,0]$. If $\phi \in C([-q,0];X)$, set $B(\phi,r) = \{g \in C([-q,0];X);\ \|g-\phi\| \leq r\}$. As usual, by $B(\phi(0),r)$ we mean the closed ball in X of radius $r > 0$ about $\phi(0)$. Let $F:]a,b[\times C([-q,0];X) \to X$ be a continuous function and D a locally closed subset of X. We will study the following semilinear initial value problem of retarded type:

$$u'(t) = Au(t) + F(t,u_t),\ u_{t_0} = \phi,\ u(t) \in D,\ t_0 \leq t \leq t_0+T \tag{2.1}$$

where $\phi \in C([t_0,\bar{T}];X)$, with $\phi(0) \in D$, $\bar{T} = t_0 + T$, $T = T(t_0,\phi) < b-t_0$. By analogy with Definition 1.3, a function u is said to be a mild solution to (2.1) on $[t_0-q,t_0+T](t_0 \in]a,b[)$ if u is continuous on $[t_0-q,t_0+T]$ and satisfies

$$u(t) = S(t-t_0)\phi(0) + \int_{t_0}^{t} S(t-s)F(s,u_s)ds,\ u_{t_0} = \phi,\ u(t) \in D \tag{2.2}$$

for all $t \in [t_0,t_0+T]$.

In the sequel, the following tangential condition will play a crucial role:

$$\lim_{h\downarrow 0} h^{-1}\, d(S(h)v(0) + hF(t,v);D) = 0,\ \forall t \in]a,b[,\ v \in C([-q,0];X). \tag{2.3}$$

with $v(0) \in D$.

By standard arguments (see (C4)' in the previous section), Condition (2.3) is equivalent to

$$\lim_{h\downarrow 0} h^{-1}\, d(S(h)v(0) + \int_{t}^{t+h} S(t+h-s)F(t,v)ds;D) = 0 \tag{2.3'}$$

for all $t \in]a,b[$ and $v \in C([-q,0];X)$ with $v(0) \in D$.

Since F is continuous at (t_0,ϕ) and D is locally closed, there are $r,T,M > 0$ with the properties

$$\|F(t,v)\| \leq M,\ \forall t \in [t_0,t_0+T],\ v \in B(\phi,r),\ t_0+T = \bar{T} < b \tag{2.4}$$

and $D \cap B(\phi(0),r)$ is closed.

Moreover, on the basis of the continuity of ϕ on $[-q,0]$, we may assume that T is sufficiently small such that (for a fixed $r_1 \in]0,r[$):

$$\|\phi(\theta_1) - \phi(\theta_2)\| \leq r-r_1,\ \forall \theta_1,\theta_2 \in]-q,0[,\ |\theta_1-\theta_2| \leq T \tag{2.5}$$

$$\max_{0 \le t \le T} \|S(t)\phi(0) - \phi(0)\| + TN(M+1) \le r_1, \ (N = C\, e^{\omega T}). \qquad (2.6)$$

For simplicity of notation, suppose that $N \le 1$. We are now able to state the main result of this section:

Theorem 2.1 *Let* $D \subset X$ *be a locally closed subset of the general Banach space* X, $F:]a,b[\times C([-q,0];X) \to X$ *a continuous function and let* A *be the infinitesimal generator of the compact* C_0 *semigroup* $S(t)$. *Then a necessary and sufficient condition for Equation* (2.1) *to have a local mild solution* u *on* $[t_0-q, t_0+T]$ *with* $T = T(t_0,\phi) > 0$, $T < b-t_0$, *for every* $t_0 \in]a,b[$ *and* $\phi \in C([-q,0];X)$, *is given by* (2.3).

Proof *The necessity of* (2.3). Let $t_0 \in]a,b[$ and $v \in C([-q,0];X)$ with $v(0) \in D$. By hypothesis, there exist $T = T(t_0,v) > 0$ with $T < b-t_0$ and a continuous function u satisfying (2.2) with $\phi = v$. Since $u(t_0+h) \in D$ for all $h \in [0,T]$, we have

$$h^{-1}\, d(S(h)v(0) + hF(t_0,v);D) \le h^{-1}\, \|S(h)v(0) + hF(t_0,v) - u(t_0+h)\|$$

$$= \|F(t_0,v) - h^{-1} \int_{t_0}^{t_0+h} S(t_0+h-s)F(s,u_s)ds\| . \qquad (2.7)$$

Letting $h \downarrow 0$, one obtains (2.3).

The sufficiency of (2.3). Given $t_0 \in]a,b[$ and $\phi \in C(-q,0];X)$, let r, r_1, T, $M > 0$ satisfy (2.4) - (2.6). We first prove that, for each positive integer n, there exist a strictly increasing sequence $\{t_i^n\}_{i=0}^{\infty} \subset [t_0,t_0+T]$ and an approximate solution u^n on $[t_0,t_0+T]$ in the following sense:

(i) $t_0^n \ \ t_0$, $t_{i+1}^n - t_i^n < \frac{1}{n}$, $\lim_{i\to\infty} t_i^n = t_0+T$.

(ii) $u_{t_0}^n = \phi$, $u^n(t_i) \in D \cap B(\phi(0),r)$, $u_{t_i}^n(0) = x_i^n \in D$, where $t_i = t_i^n$.

(iii) $u^n(t) = S(t-t_i^n)x_i^n + \int_{t_i^n}^{t} S(t-s)F(t_i^n,u_{t_i}^n)ds + (t-t_i^n)p_i^n$

for $t \in [t_i^n,t_{i+1}^n]$, where $x_i^n = u^n(t_i^n)$ and $\|p_i^n\| \le 1/n$. Moreover, $u_{t_i}^n \in B(\phi,r) \cap C([-q,0];X)$.

Since there is no danger of confusion, we drop n as a superscript for t_i, x_i, u, p_i etc. The construction of u and t_i is by induction. Set

$t_0^n \equiv t_0$, $u_{t_0} = \phi$, $u(t_0) = \phi(0) \equiv x_0$ and suppose that u is constructed on $[t_0-q,t_i]$. Then define t_{i+1} in the following manner: if $t_i = t_0 + T$, set $t_{i+1} = t_0 + T$, and if $t_i < t_0 + T$, then we proceed as below. Set

$$\delta_i = \sup \{h \in]0,1/n];\ t_i + h \leq t_0 + T,$$

$$d(S(h)x_i + \int_{t_i}^{t_i+h} S(t_i+h-s)F(t_i,u_{t_i})ds;D) \leq h/2n\}. \tag{2.8}$$

Clearly (2.3)', along with the induction hypotheses, assures that $\delta_i > 0$. Choose a number $d_i \in]2^{-1}\delta_i,\delta_i]$ (hence $0 < d_i \leq 1/n$) such that

$$d(S(d_i)x_i + \int_{t_i}^{t_i+d_i} S(t_i+d_i-s)F(t_i,u_{t_i})ds;D) \leq d_i/2n. \tag{2.9}$$

Define $t_{i+1} = t_i + d_i$. By (2.9) there is a member x_{i+1} of D such that

$$\|S(d_i)x_i + \int_{t_i}^{t_{i+1}} S(t_{i+1}-s)F(t_i,u_{t_i})ds - x_{i+1}\| \leq d_i/n.$$

Consequently, x_{i+1} can be written in the form

$$x_{i+1} = S(t_{i+1}-t_i)x_i + \int_{t_i}^{t_{i+1}} S(t_{i+1}-s)F(t_i,u_{t_i}) + (t_{i+1}-t_i)p_i \tag{2.10}$$

with $\|p_i\| \leq 1/n$. We now define u on $[t_i,t_{i+1}]$ as indicated by (iii).

Let us check that $u_{t_{i+1}} \in B(\phi,r)$. To do this, we first note that u can be rewritten as

$$u(t) = S(t-t_0)\phi(0) + \sum_{j=0}^{i-1} \int_{t_j}^{t_{j+1}} S(t-s)F(t_j,u_{t_j})ds + \int_{t_i}^{t} S(t-s)F(t_i,u_{t_i})ds$$

$$+ \sum_{j=0}^{i-1} (t_{j+1}-t_j)S(t-t_{j+1})p_j+(t-t_i)p_i,\ t_i \leq t \leq t_{i+1}. \tag{2.11}$$

This elementary fact can be proved by induction on i (which is left to the reader). We have to estimate $\|u_{t_{i+1}}(\theta) - \phi(\theta)\|$ for each $\theta \in [-q,0]$. If $-q \leq \theta \leq t_0 - t_{i+1}$, then by (2.5)

$$\|u_{t_{i+1}}(\theta) - \phi(\theta)\| = \|u(t_o + (t_{i+1} + \theta - t_o)) - \phi(\theta)\|$$

$$= \|\phi(t_{i+1} + \theta - t_o) - \phi(\theta)\| < r - r_1 < r$$

since $t_{i+1} - t_o < T$. If $t_o - t_{i+1} \leq \theta \leq 0$, then $t_{i+1} + \theta \geq t_o$ so $|\theta| \leq t_{i+1} - t_o \leq T$. Hence, (2.11), (2.6) and the induction hypothesis yield

$$\|u_{t_{i+1}}(\theta)-\phi(\theta)\| \leq \|u(t_{i+1}+\theta)-\phi(0)\| + \|\phi(0)-\phi(\theta)\|$$

$$\leq \|S(t_{i+1}+\theta-t_o)\phi(0)-\phi(0)\| + T(M+1) + \|\phi(0)-\phi(\theta)\|$$

$$\leq r_1 + r - r_1 = r$$

and hence $u_{t_{i+1}} \in B(\phi,r)$. Using once again (2.11), we derive (for $t \in [t_o, t_{i+1}]$),

$$\|u(t) - \phi(0)\| \leq \|S(t-t_o)\phi(0) - \phi(0)\| + T(M+1) \leq r_1 < r$$

i.e., $u(t) \in B(\phi(0),r)$, for $t \in [t_o, t_{i+1}]$. This remark, along with the fact that $\phi \in C([-q,0],X)$, implies that $u_{t_{i+1}} \in B(\phi,r) \cap C([-q,0];X)$.

If $t^* = \lim_{i\to\infty} t_i$, then $t^* \leq t_o + T$. We have to prove that $t^* = t_o + T$. For this purpose we first prove that $\lim_{i\to\infty} x_i = x^*$ also exists. Indeed, let $j > i$. Using (2.11) for $t = t_i$ and $t = t_j$, we derive

$$\|x_i - x_j\| \leq \|S(t_j-t_i)\phi(0) - \phi(0)\|$$
$$+ \sum_{m=0}^{i-1} \|S(t_j-t_i)-I)F(t_m,u_{t_m})\| (t_{m+1}-t_m)$$
$$+ \sum_{m=0}^{i-1} \|S(t_j-t_i)p_m - p_m\| (t_{m+1}-t_m) + (M+1)(t_j-t_i). \qquad (2.12)$$

We now proceed, exactly as in the proof of the convergence of x_i given by (1.12), to show that (2.12) implies that x_i given by (2.10) is also a Cauchy sequence. Consequently, $\lim_{i\to\infty} x_i = x^*$ exists. Since $x_i \in D \cap B(\phi(0),r)$ (which is closed), $x^* \in D \cap B(\phi(0),r)$. Define $u(t^*) = x^*$. By (iii) we have

$$\|u(t)-x_i\| = \|u(t)-u(t_i)\| \leq \|S(t-t_i)x_i-x_i\| + (t-t_i)(M+1)$$

and therefore $\lim_{t\uparrow t^*} u(t) = x^* = u(t^*)$. Accordingly, u is continuous on $[t_o-q,t^*]$ and hence $\lim_{i\to\infty} u_{t_i} = u_{t^*} \in C([-q,0];X)$. This shows also that $u_{t^*} \in B(\phi,r) \cap C([-q,0];X)$. We now assume by contradiction that $t^* < t_o + T$. In view of (2.3)' we may choose $h^* \in]0,1/n]$ sufficiently small so that

$$d(S(h^*)x^* + \int_{t^*}^{t^*+h^*} S(t^*+h^*-s)F(t^*,u_{t^*})ds;D) \leq \frac{h^*}{4n}. \tag{2.13}$$

Since $2^{-1}\delta_i < d_i$ and $d_i = t_{i+1} - t_i \to 0$ as $i \to \infty$, there is a positive integer i_o such that $\delta_i < h^*$ for all $i \geq i_o$. On the basis of (2.3) we have

$$d(S(h^*)x_i + \int_{t_i}^{t_i+h} S(t_i+h^*-s)F(t_i,u_{t_i})ds;D) > \frac{h^*}{2n} \tag{2.14}$$

for all $i \geq i_o$. Letting $i \to \infty$ in (2.14), one obtains an inequality which contradicts (2.13) and we conclude that $t^* = t_o+T$. To continue, let us introduce the following functions: $a_n(s) = t_i$ for $s \in [t_i,t_{i+1}[$, $a_n(t_o+T) = t_o+T$ and

$$g_n(t) = \sum_{j=0}^{i-1} (t_{j+1}-t_j)S(t-t_{j+1})p_j+(t-t_i)p_i,\ t_i \leq t \leq t_{i+1}.$$

It is now clear that u^n (given by (iii)) can be written in the form

$$u^n(t) = S(t-t_o)\,\phi(0) + \int_{t_o}^{t} S(t-s)F(a_n(s),u^n_{a_n(s)})ds + g_n(t) \tag{2.15}$$

for all $t \in [t_o,t_o+T]$, $u^n_{t_o} = \phi$. Set

$$y^n(t) = \int_{t_o}^{t} S(t-s)F(a_n(s),\ u^n_{a_n(s)})ds,\ t \in [t_o,t_o + T]. \tag{2.16}$$

Let $t > t_o$ and $\varepsilon \in]0,t-t_o[$. Define

$$y^n_\varepsilon(t) = \int_{t_o}^{t-\varepsilon} S(t-s)F(a_n(s),u^n_{a_n(s)})ds.$$

Clearly,

$$y^n_\varepsilon(t) = S(\varepsilon) \int_{t_o}^{t-\varepsilon} S(t-s-\varepsilon)F(a_n(s),u^n_{a_n(s)})ds.$$

Once this is accomplished, the rest of the proof of the equicontinuity of $t \to y^n(t)$ and of precompactness of $\{y^n(t)\}_{n=1}^{\infty}$ is exactly as in the proof of Theorem 1.1 (so we omit it). Consequently we may assume that $\lim_{n\to\infty} y^n = y$ exists in $C([-q,0];X)$. Since $\|g_n(t)\| \leq T/n$, for all $t \in [t_0,t_0+T]$ it follows that $\lim_{n\to\infty} u^n(t) = S(t-t_0)\phi(0) + y(t) \equiv u(t)$ uniformly with respect to $t \in [t_0,t_0+T]$. Let us observe that $(a_n(s),u^n_{a_n(s)}) \to (s,u_s)$ as $n \to \infty$. Moreover $u^n(a_n(s)) \in D \cap B(\phi(0),r)$ (which is supposed to be closed), implies $u(s) \in D \cap B(\phi(0),r)$. Finally, passing to the limit in (2.15) one obtains (2.2). The proof is complete.

Remark 2.1 If, in addition to the hypotheses of Theorem 2.1, we suppose that $\phi(\theta) \in D$ for all $\theta \in [-q,0]$, then there exists a solution to (2.2) with $u(t) \in D$ for all $t \in [t_0-q,t_0+T]$.

Let us now proceed to discuss briefly the semilinear differential equation with retarded argument

$$u'(t) = Au(t) + f(t,u(t-q)),\ u(t) \in D,\ t_0 \leq t \leq t_0 + T \tag{2.17}$$

with the initial condition

$$u_{t_0} = \phi,\ \phi \in C([-q,0];D) \tag{2.18}$$

where $C([-q,0];D)$ denotes the set of all continuous functions from $[-q,0]$ into D.

By a mild solution u to (2.17) on $[t_0-q,t_0+T]$ we mean a continuous $u:[t_0-q,t_0+T] \to D$ satisfying

$$u(t) = S(t-t_0)\phi(0) + \int_{t_0}^{t} S(t-s)f(s,u(s-q))ds,\ t \in [t_0,t_0+T] \tag{2.19}$$

and the initial condition (2.18), i.e., $u(t_0+\theta) = \phi(\theta)$, $\theta \in [-q,0]$. Here $f:]a,b[\times X \to X$ is supposed to be continuous. In this case, D is said to be a flow-invariant set with respect to (2.17), if for each $t_0 \in]a,b[$ and $\phi \in C([-q,0];D)$ Equation (2.19) has a D-valued solution. We are not going to characterize the flow-invariance of D with respect to (2.17). We restrict ourselves to the following local existence result:

Theorem 2.2 *Let* D *be a locally closed subset of* X *and let* A *be the infinitesimal generator of the compact* C_0 *semigroup* $S(t)$. *Moreover, assume that* $f:]a,b[\times X \to X$ *is continuous and satisfies the condition*

$$\lim_{h\downarrow 0} h^{-1} d(S(h)v(0) + hf(t,v(-q));D) = 0 \tag{2.20}$$

for all $t \in]a,b[$ *and* $v \in C([-q,0];X)$, *with* $v(0), v(-q) \in D$. *Then for each* $t_0 \in]a,b[$ *and* $\phi \in C([-q,0];D)$ *there exists* $T = T(t_0,\phi) > 0$ *and a continuous function* $u:[t_0,t_0+T] \to D$ *satisfying* (2.19).

Proof Let $t_0 \in]a,b[$ and $\phi \in C([-q,0];D)$. There exists $M,r,T > 0$ such that

$$\|f(t,y)\| \leq M, \ \forall t \in [t_0,t_0+T], \ x \in D \cap B(\phi(0),r), \ 0 < T < q \tag{2.21}$$

$$\max_{0\leq t\leq T} \|S(t)\phi(0) - \phi(0)\| + TN(M+1) \leq r, \ N = Ce^{\omega T}. \tag{2.22}$$

We may also assume that $D \cap B(\phi(0),r)$ is closed and $\|f(t,\phi(\theta))\| \leq M$, $\forall \theta \in [-q,0]$.

The rest of the proof is similar to that of Theorem 2.1. We have to observe that in this case

$$u^n(t) = S(t-t_i)x_i + \int_{t_i}^{t} S(t-s)f(t_i,u^n(t_i-q))ds + (t-t_i)p_i$$

for all $t \in [t_i,t_{i+1}]$, with $u^n_{t_0} = \phi$ and $\|p_i\| \leq 1/n$. Moreover, $u_{t_i} \in C([-q,0];X)$ and

$$u_{t_i}(-q) = u(t_i-q) = u(t_0+t_i-t_0-q) = \phi(t_i-t_0-q) \in D$$

since $t_i \leq t_0 + T \leq t_0 + q$. According to (2.20) there is $h \in]0,1/n]$ such that $t_i + h \leq t_0 + T$ and

$$d(S(h)x_i + \int_{t_i}^{t_i+h} S(t_i+h-s)f(t_i,u(t_i-q))ds;D) \leq h/2n$$

and so on. The conclusion of the theorem follows by passing to the limit for $n \to \infty$ in the formula

$$u^n(t) = S(t-t_0)\phi(0) + \int_{t_0}^{t} S(t-s)f(a_n(s),u^n(a_n(s)-q))ds + g_n(t)$$

$$t \in [t_0,t_0 + T].$$

Remark 2.2 If we are interested in the condition "$\phi(0) \in D$" rather than "$\phi(\theta) \in D$ for all $\theta \in [-q,0]$", then Theorem 2.2 can be restated as:

Theorem 2.3 *Let* D *be locally closed and let* A *be the infinitesimal generator of the compact* C_0 *semigroup* S(t). *A necessary and sufficient condition for Equation* (2.17) *to have a local mild solution* $u:[t_0,t_0+T] \to D$, *with* $T = T(t_0,\phi) > 0$, (*for every* $t_0 \in]a,b[$ *and* $\phi \in C([-q,0];X)$, *with* $\phi(0) \in D$) *is given by*

$$\lim_{h\downarrow 0} h^{-1} d(S(h)v(0) + hf(t,v(-q));D) = 0 \qquad (2.23)$$

for all $t \in]a,b[$ *and* $v \in C([-q,0];X)$ *with* $v(0) \in D$.

The proof of the necessity of (2.23) is exactly the proof of the necessity of (2.3), with $f(s,u(s-q))$ in place of $F(s,u_s)$.

In the case in which D is open, the tangential conditions (2.3) and (2.20) are automatically satisfied. Moreover, if X is finite dimensional (say, $X = R^n$) then we can take $A = 0$. In this case, by Theorem 2.1 one obtains the well-known result of J.K. Hale [1]:

Theorem 2.3 (Hale). *Let* $F:]a,b[\times C([-q,0];R^n) \to R^n$ *be a continuous function. Then for each* $t_0 \in]a,b[$ *and* $\phi \in C([-q,0];R^n)$ *the problem*

$$u'(t) = F(t,u_t),\ u_{t_0} = \phi,\ t \in [t_0,t_0+T] \qquad (2.24)$$

has a (local) solution, for some $T = T(t_0,\phi) > 0$, *with* $T < b - t_0$.

§3. Semilinear equations with dissipative time-dependent domain perturbations

Throughout this section, X is a real Banach space. We will consider again Problem (1.30), rewritten here as

$$u'(t) = Au(t) + F(t,u(t)),\ u(t_0) = x \in D(t_0),\ u(t) \in D(t) \qquad (3.1)$$

for all $t \in [t_0,b[$, where $t_0 \in]a,b[$.

Now the conditions are different from those in §1. First of all we drop the hypothesis of the compactness of S(t), $t > 0$. In this case (i.e., if we assume only that A is the infinitesimal generator of the C_0 semigroup and F is continuous), (3.1) may have no solution, even if $A = 0$. (See Godunov's theorem in Chapter 2, §1.) Secondly, in addition to the continuity of F, we add the g-dissipativity (i.e., Hypothesis (D4) below), which is strictly more general than the Lipschitz condition.

Let us now proceed to give our hypotheses in a precise form:

(D1) $A:D(A) \subset X \to X$ *is the infinitesimal generator of the* C_0 *semigroup* $S(t)$, $t \geq 0$, *of type* $\omega \geq 0$, *i.e.,* $\|S(t)\| \leq e^{\omega t}$, $t \geq 0$.

(D2) *The multivalued mapping* $t \to D(t) \equiv D(F(t,\cdot))$ *satisfies Condition* (C2) *in* §1.

(D3) *The function* $(t,x) \to F(t,x)$ *is continuous at every* (t,x) *with* $t \in]a,b[$ *and* $x \in D(t)$.

(D4) *For each* $t \in]a,b[$, *the operator* $x \to F(t,x)$ *from* $D(t)$ *into* X, *is* $g(t)$*-dissipative, that is*

$$(1-\lambda g(t)) \|x_1-x_2\| \leq \|x_1-x_2 - \lambda(F(t,x_1) - F(t,x_2))\|$$

for all $x_1,x_2 \in D(t)$ *and* $\lambda > 0$, *with* $g:]a,b[\to [0, +\infty[$ *nondecreasing.*

(D5) $\lim_{h\downarrow 0} h^{-1} d(S(h)x + hF(t,x);D(t+h)) = 0, \quad \forall t \in]a,b[, \ x \in D(t).$

Condition (D4) *is equivalent to:*

(D4)' $\langle F(t,x_1) - F(t,x_2,x_1-x_2\rangle_i \leq g(t) \|x_1-x_2\|^2$

for all $t \in]a,b[$ *and* $x_1,x_2 \in D(t)$. (*See* (3.3) *and* (3.4) *in Chapter* 1.)

Condition (D5) *is equivalent to:*

(D5)' $\lim_{h\downarrow 0} h^{-1} d(S(h)x + \int_t^{t+h} S(t+h-s)F(t,x)ds;D(t+h)) = 0,$

for all $t \in]a,b[$ *and* $x \in D(t)$.

Finally, if in (D5) *we replace* "lim" *by* "lim inf", *one also obtains an equivalent condition.*

We shall see that (D1) - (D5) guarantee the local existence (and uniqueness) of the mild solution u to (3.1). As in the previous section, by a mild solution u on $J_u \subset [t_0,b[$ with $t_0 \in J_u$, to (3.1), we mean a continuous function u from J_u into X, satisfying

$$u(t) = S(t-t_0)x + \int_{t_0}^{t} S(t-s)F(s,u(s))ds, \ u(t) \in D(t), \forall t \in J_u \qquad (3.2)$$

However, for global existence additional hypotheses are needed, namely:

(E1) *The mapping* $t \to D(t)$ *from* $]a,b[$ *into* 2^X *(the family of all subsets of* X*) is closed.*

(E2) *For each* $s \in]a,b[$ *and for each connected component* $C(s)$ *of* $D(s)$*, there exists a continuous function* $w:[s,\bar{b}] \to X$ *such that* $w(s) \in D(s)$ *and* $w(t) \in D(t)$, $\forall t \in [s,\bar{b}[$ *(for some* $\bar{b} \in]a,b]$*).*

We are now prepared to state the main result of this section:

Theorem 3.1 *Suppose that* (D1) - (D4) *are fulfilled. Then a necessary and sufficient condition for Problem* (3.1) *to have a local (unique, in view of* (D4)*) mild solution* u *on* $[t_o,t_o+T]$, $0 < T = T(t_o,x)$, *(for every* $t_o \in]a,b[$ *and* $x \in D(t_o)$*) is given by* (D5).

Sketch of the proof The complete proof of Theorem 3.1 is quite long and difficult. It is given in the author's paper [15]. Here we point out the main steps of this proof, only. One first proves the existence of an approximate solution of the form (a3) in the proof of Proposition 1.1, that is,

$$u_n(t) = S(t-t_i)x_i + \int_{t_i}^{t} S(t-s)F(t_i,x_i)ds + (t-t_i)p_i \tag{3.3}$$

for all $t \in [t_i,t_{i+1}]$, where $\|p_i\| \leq 1/n$, $u_n(t) \in B(x,r)$. If $t_i = t_o + T$, set $t_{i+1} = t_i$. If $t_i < t_o + T$, set $t_{i+1} = t_i + d_i$ where d_i is the maximal number in $]0,1/n]$ with the properties:

$$\|F(t,y) - F(t_i,x_i)\| \leq 1/n, \ \forall t \in [t_i,t_i+d_i], \ y \in D(t),$$

with

$$\|y - x_i\| \leq \sup_{0\leq s\leq d_i} \|S(s)x_i-x_i\| + d_i(M+1)N, \ N = e^{\omega T} \tag{3.4}$$

$$\sup_{0\leq s\leq d_i} \|S(s)z-z\| \leq 1/n, \ \forall z \in \{x_j, F(t_j,x_j), \ j = 0,1,\ldots,i\} \tag{3.5}$$

$$d(S(d_i)x_i + \int_{t_i}^{t_i+d_i} S(t_i+d_i-s)F(t_i,x_i)ds;D(t_i+d_i)) \leq \frac{d_i}{2n}. \tag{3.6}$$

On the basis of (3.6), there exists $x_{i+1} \in D(t_i+d_i)$ of the form (1.12) with $F(t_i,x_i)$ in place of $f(t_i,x_i)$. In such a manner we construct the sequence u_n on $[t_o,t_o+T]$ of the form (1.17), that is

$$u_n(t) = S(t-t_o)x + \int_{t_o}^{t} S(t-s)F(a_n(s), u_n(a_n(s)))ds + g_n(t) \qquad (3.7)$$

with $u_n(t_i) = x_i \in D(t_i) \cap B(x,r)$.

Up to this point the proof is very similar to that of Proposition 1.1. From now on, the proof differs. This is because the convergence of u_n is not at all obvious. To prove that the g-dissipativity of F assures the convergence of u_n, one proceeds as follows: let m and n be positive integers. Set $U = \{t_i^n, t_j^m, i,j = 0,1,...\}$. Denote by $\{r_e\}_{e=0}^{\infty}$ the minimal refinement at partitions $\{t_i^n\}$ and $\{t_j^m\}$ of $[t_o,t_o+T]$, that is

$$r_o = t_o, \; r_{e+1} = \min \{t \in U;\; t > r_e\},\; e = 0,1,... .$$

For each positive integer e, one constructs a partition $\{s_k^e\}_{k=0}^{\infty}$ of $[r_e,r_{e+1}]$ and (similarly to u_n) the functions v^n and v^m given by

$$v^p(t) = S(t-s_k)v_k^p + \int_{s_k}^{t} S(t-s)F(s_k,v_k^p)ds + (t-s_k)q_k^{p,e} \qquad (3.8)$$

with $\|q_k^{p,e}\| \leq 1/p$, $r_e \leq s_k \leq t < s_{k+1} \leq r_{e+1}$, $p = n,m$ and $k = 0,1,...$. Here, $s_o = r_e$, $v^p(t_o) = x$ and $0 < s_{k+1} - s_k \equiv h_k$ with $h_k = r_{e+1}-r_e \leq \min \{1/n, 1/n\}$, $\lim_{k\to\infty} s_k = r_{e+1}$.

The elements $v_k^p \in D(s_k) \cap B(x,r)$ are constructed similarly to x_i^n. We have to mention that if we are in the situation $t_i^n, t_j^m \leq r_e < r_{e+1} \leq t_{i+1}^n, t_{j+1}^m$, then we define

$$v^n(r_e) = u_n(r_e) \text{ if } r_e = t_i^n, \text{ and } v^n(r_e) = v^n(r_e-) \text{ if } r_e = t_j^m;$$

$$v^m(r_e) = u_m(r_e) \text{ if } r_e = t_j^m, \text{ and } v^m(r_e) = v^m(r_e-) \text{ if } r_e = t_i^n.$$

These sequences v^p have the following useful properties:

$$\left.\begin{aligned} \|v^n(t) - u_n(t)\| &\leq 3(t-t_i^n)N/n \leq 3TN/n \\ \|v^m(t) - u_m(t)\| &\leq 3(t-t_j^m)N/m \leq 3TN/m \end{aligned}\right\} \quad t \in [r_e,r_{e+1}] \qquad (3.9)$$

for the above situation. Moreover, by using g-dissipativity of F one proves that

$$\|v^n(r_e-)-v^m(r_e-)\| \leq 6NT(m^{-1} + n^{-1}) \exp (2g(\bar{T})T + \omega T) \qquad (3.10)$$

where $\bar{T} = t_0 + T$ and $v^p(r_e-) = \lim_{t \uparrow r_e} v^p(t)$, $e = 0,1,\ldots$.

On the basis of (3.10) we can prove that $v^n(t)$ is uniformly Cauchy on $[t_0,t_0+T]$. Then (3.9) implies that $u_n(t)$ is also uniformly Cauchy on $[t_0,t_0+T]$. Consequently, $\lim_{n\to\infty} u_n(t) \equiv u(t)$ exists. Arguing exactly as in the proof of Theorem 1.4, it follows that u is a solution to (3.2) on $[t_0,t_0+T]$ and that (D5) is also necessary.

The uniqueness. In general, a mild solution to (3.1) may not be differentiable (see the remark following Definition 1.4). Consequently, for proof of uniqueness we cannot use the standard method involving Kato's lemma, as in the proof of Inequality (2.30) in Chapter 2.

There are several methods for proving uniqueness in the nondifferentiable case. One is based on differential inequalities, as follows: let u_1 and u_2 be solutions to (3.2) on $J = [t_0,t_0+T] \subset [t_0,b[$ with $u_1(t_0) = x$, $u_2(t_0) = y$, $x, y \in D(t_0)$, $t_0 \in]a,b[$. Using g-dissipativity of F one proves (see, e.g., the author [15])

$$D_- \|u_1(t) - u_2(t)\| \leq (g(t) + \omega) \|u_1(t) - u_2(t)\|, \quad t \in J \tag{3.11}$$

where $D_-f(t) = \limsup_{h\downarrow 0} (f(t-h) - f(t))/(-h)$. Solving the differential inequality (3.11), one obtains

$$\|u_1(t) - u_2(t)\| \leq (\|x-y\|) \exp(g(t) + \omega)(t - t_0), \quad t \in J \tag{3.12}$$

which implies the uniqueness of the mild solution to (3.1).

Another method is based on Benilan's inequality in the theory of integral solutions for nonlinear differential equations (of the form (3.1) with A nonlinear (possible) multivalued and dissipative). (See Benilan [1], Barbu [2] or the author [14].)

In the sequel we give an elementary technique (for uniqueness) which is independent of both differential inequalities and integral solutions. It is also interesting in itself. Since u_1 and u_2 satisfy (3.2), we have

$$u_i(t+h) = S(h)u_i(t) + \int_t^{t+h} S(t+h-s)F(s,u_i(s))ds, \quad t \in J;\ i = 1,2 \tag{3.13}$$

for all $h > 0$ with $t + h \in J$.

On the other hand, (D4) yields

$$(1-g(t+h)h)\,\|u_1(t+h) - u_2(t+h)\|$$

$$\leq \|u_1(t+h)-u_2(t+h)-hF(t+h,u_1(t+h))+hF(t+h,u_2(t+h)\|. \qquad (3.14)$$

Combining (D1), (3.13), (3.14) and the inequality

$$(1 - h)^{-1} \leq e^{(1+\varepsilon)h}, \ \forall h \in]0,\varepsilon/(1+\varepsilon)], \ \varepsilon > 0 \qquad (3.15)$$

we get

$$\|u_1(t+h)-u_2(t+h)\| \leq (\| u_1(t)-u_2(t)\| + \|\bar{I}\|)\exp(1+\varepsilon)g(t+h)+\omega)h \qquad (3.16)$$

where h is small (such that $hg(\bar{T})(1+\varepsilon) \leq \varepsilon$, $\bar{T} = t_0 + T$) and

$$\bar{I} = \int_t^{t+h} S(t+h-s)(F(s,u_1(s)) - F(s,u_2(s)))$$

$$- (F(t+h,u_1(t+h)) - F(t+h,u_2(t+h)))ds.$$

Set $f(t) = F(t,u_1(t)) - F(t,u_2(t))$. Let $K = \{f(t), t \in [t_0,t_0+T]\}$. There exists $d = d(\varepsilon) > 0$ such that

$$\|S(s)z-w\| \leq \varepsilon, \ \forall s \in]0,d[, \ \|z-w\| \leq d, \ \forall z,w \in K. \qquad (3.17)$$

Choose $r = r(\varepsilon) > 0$ with the property that

$$\|f(t) - f(s)\| \leq d, \ \forall t, \ s \in [t_0,t_0+T], \ |t-s| \leq r.$$

For $s \in [t,t+h]$, we have $0 \leq t+h-s \leq h$, hence

$$\|f(t+h)-f(s)\| \leq d, \ \forall s \in [t,t+h], \ 0 < h < r \qquad (3.18)$$

and therefore by (3.17) we have

$$\|\bar{I}\| \leq h\varepsilon, \ \forall h < d_0(\varepsilon) = \min \{d,r\}, \ \ \forall t \in [t_0,t_0+T], \ t+h < t_0+T.$$

Now, from (3.16) we derive

$$\|u_1(t+h)-u_2(t+h)\| \leq (\| u_1(t)-u_2(t)\| + \varepsilon h)\exp(1+\varepsilon)g(t+h)+\omega)h \qquad (3.19)$$

for all $t \in [t_0,t_0+T[$, $h \leq h_0 = \min \{d_0(\varepsilon), \varepsilon/(1+\varepsilon)g(\bar{T})\}$. To obtain (3.12), fix $t \in [t_0,t_0 + T]$ and choose a partition $t_0 < t_1 < \ldots < t_n = t$ of $[t_0,t]$, with $t_k-t_{k-1} = h_k \leq h_0$. Then (3.19) yields

$$\|u_1(t_k)-u_2(t_k)\| \leq (\| u_1(t_{k-1})-u_2(t_{k-1})\| + \varepsilon h_k)\exp(1+\varepsilon)g(t_k) + \omega)h_k.$$

Iterating this inequality for k = 1,2,...,n, one obtains

$$\|u_1(t)-u_2(t)\| \leq (\| x-y\| + (t-t_0)\varepsilon) \exp((1+\varepsilon)g(t)+\omega)(t-t_0) \qquad (3.20)$$

for every $\varepsilon > 0$, which is equivalent to (3.12). This completes the proof of uniqueness.

§4. The extendability of mild solutions

This section is devoted to the problem of continuation of the local solution to the equations in previous sections. More precisely, one considers the initial value problem

$$u(t) = S(t-t_0)x + \int_{t_0}^{s} S(t-s)F(s,u(s))ds, \ t \in [t_0,\bar{b}[\qquad (4.1)$$

with $a < t_0 < \bar{b} \leq b$, $x \in D(t_0)$, under Hypotheses (D1)-(D5) and (E1)-(E2) in the previous section. Condition (C1) in Section 1 will be also considered. The concepts of continuation and noncontinuability of the solution to (4.1) are defined as in Chapter 2, §1, §2.

A basic property of the solutions to (4.1) is the following:

Proposition 4.1 *Let* u *and* v *be solutions* (4.1) *on* $[t_0,T]$ *and* $[T,T_1]$ *respectively, with* $t_0 < T < T_1$ *and* $u(T) = v(T)$. *Define* $w:[t_0,T_1] \in X$ *by* $w(t) = u(t)$ *on* $[t_0,T]$ *and* $w(t) = v(t)$ *on* $[T,T_1]$. *Then* w *is a solution to* (4.1) *on* $[t_0,T_1]$.

The proof is routine so we omit it.

We can now prove the following result:

Proposition 4.2 *Suppose that* (D1)-(D3) *and* (D5) *are fulfilled and that for each* $t_0 \in]a,b[$ *and* $x \in D(t_0)$ *Problem* (4.1) *has a local solution* u *on* $[t_0,t_0+T]$ *with* $T = T(t_0,x) > 0$. *Suppose in addition that* $t \to D(t)$ *is closed on* $]a,b[$. *Then each solution* u *to* (4.1) *has a noncontinuable continuation* v, *which is defined on an interval of the form* $[t_0,\bar{b}[$ *with* $t_0 < \bar{b} \leq b$. *If* $\bar{b} < b$, *then* $\lim_{t\uparrow b} v(t)$ *does not exist.*

Proof For the first part of the conclusion one proceeds as in the proof of Proposition 2.2 in Chapter 2. The proof of the second part is by contradiction. Therefore, suppose that $\bar{b} < b$ and that $\lim_{t\uparrow\bar{b}} v(t) = \bar{v}$ exists. Since $v(t) \in D(t)$, it follows that $\bar{v} \in D(\bar{b})$. By standard arguments involving Proposition 4.1, we conclude that v can be extended to the right of $\bar{b}$, which contradicts the noncontinuability of v.

We are now prepared to prove:

Proposition 4.3 *In addition to the hypotheses of Proposition* 4.2, *suppose that* F *maps bounded subsets in* $]a,b[\times X$ *into bounded subsets of* X. *Let* u *be a noncontinuable solution to* (4.1) *on* $[t_0,b_0[$. *Then either* $b_0 = b$, *or* (*if* $b_0 < b$) $\lim_{t\uparrow b} \|u(t)\| = +\infty$.

Proof We first prove that if $b_0 < b$ then u is unbounded on $[t_0,b_0[$, that is, $\overline{\lim_{t\uparrow b_0}} \|u(t)\| = +\infty$. Indeed, if we assume by contradiction that u is bounded on $[t_0,b_0[$, then by our assumption it follows that there exists a constant $Q > 0$ such that $\|F(t,u(t))\| \leq Q$, for all $t \in [t_0,b_0[$. This will lead to the existence of $\lim_{t\uparrow b_0} u(t)$, which is a contradiction (in view of Proposition 4.2). To prove the existence of the previous limit, take $t_0 = 0$ (to simplify notation) and $0 < \varepsilon < t < t' < b_0$. Then by (4.1) we derive (with $N = S(t)$, $t \in [t_0,b_0[$).

$$\|u(t') - u(t)\| \leq \|S(t')x - S(t)x\| + (t'-t)QN \left\| \int_0^t (S(t'-s) - S(t-s))F(s,u(s))ds \right\| . \tag{4.2}$$

It is easy to see that the last term I_3 in the left-hand side is bounded from above as follows:

$$I_3 \leq 2QN\varepsilon + N \int_0^{t-\varepsilon} \|(S(t'-t)F(s,u(s)) - F(s,u(s)))\| \, ds$$
$$\leq 2NQ\varepsilon + N \int_0^{b_0-\varepsilon} \|S(t'-t)F(s,u(s)) - F(s,u(s))\| \, ds.$$

Let us consider the compact $K_\varepsilon = \{F(s,u(s)),\ 0 \leq s \leq b_0-\varepsilon\}$. Since $(h,z) \to S(h)z$ is continuous from the compact $[0,b_0-\varepsilon] \times K_\varepsilon$ into X, there exists

$d(\varepsilon) > 0$ such that

$$\|S(t'-t)F(s,u(s))-F(s,u(s))\| \leq \varepsilon,$$

for $t'-t \leq d(\varepsilon)$ and for all $s \in [0,b_0-\varepsilon]$. Consequently,

$$I_3 \leq 2NQ\varepsilon + (b_0-\varepsilon)\varepsilon, \quad \forall t'-t < d(\varepsilon). \tag{4.3}$$

Going back to (4.2), we conclude that $\lim_{t\uparrow b_0} u(t)$ exists. This contradiction shows that u is unbounded on $[t_0,b_0[$.

To conclude the proof, it suffices to show that $\liminf_{t\uparrow b_0} \|u(t)\| = +\infty$. Suppose that this is false. Then there exists an increasing sequence $\{t_k\} \subset [t_0,b_0[$ and $r > 0$ with the properties:

$$\|u(t_k)\| \leq r, \; k = 1,2,\ldots, \; \lim_{k\to\infty} t_k = b_0.$$

As in the proof of Proposition 2.3 in Chapter 2, set

$$M = \sup \{\|F(t,y)\|, \; t \in [t_0,b_0], \; y \in D(t), \; \|y\| \leq (r+1)N\}.$$

Let $a \in [t_0,b_0[$ be such that $(b_0-a)M < 1$ and let i be a positive integer with the property that $a < t_k$ for all $k \geq i$. Fix $k = i$. Then we can prove the estimate

$$\|u(t)\| < (r+1)N, \; \forall t \in [t_k,b_0[\tag{4.4}$$

(which contradicts the unboundedness of u on $[t_0,b_0[$). Indeed, if this were false there would exist $h > 0$ such that $t_k+h < b_0$, $\|u(t_k+s)\| <(r+1)N$ for all $s \in [0,h]$, and $\|u(t_k+h)\| = N(r+1)$. On the other hand,

$$u(t_k+h) = S(h)u(t_k) + \int_{t_k}^{t_k+h} S(t_k+h-s)F(s,u(s))ds$$

yields the contradiction

$$N(r+1) = \|u(t_k+h)\| \leq Nr + hMN \leq N(r+(b_0-a)M) < N(r+1)$$

which completes the proof.

<u>Remark 4.1</u> In the previous results we did not use the compactness of S(t)

for $t > 0$. In particular, we have established (4.3) without the continuity of $S(t)$ in the uniform operator topology (at $t > 0$), as it is usually used (see, e.g., Pazy [5], p. 194). However, we have already seen that the compactness of $S(t)$ is important for local existence.

Combining Theorem 1.4 and Proposition 4.3, one obtains:

Theorem 4.1 *Suppose, in addition to the hypotheses of Theorem* 1.4, *that* $t \to D(t)$ *is closed and that* F *maps bounded subsets into bounded subsets. Then, for each* $t_0 \in]a,b[$ *and* $x \in D(t_0)$, *the initial value problem* (1.30) *has a noncontinuable mild solution* u *on (a maximum interval of existence)* $[t_0,b_0[$ $(b_0 \equiv t_{max})$. *If* $b_0 < b$, *then* $\lim_{t \uparrow b_0} \|u(t)\| = +\infty$. *Moreover, the tangential condition* (1.32) *is also necessary.*

In the case in which $D(t) = D$ is independent of t, the fact that $t \to D(t)$ is closed means that D is closed. If F is continuous on $]a,b[\times D$ we may assume, without loss of generality (by Titze's theorem), that F is continuous on $]a,b[\times X$. In other words, Theorem 4.1 is actually a characterization of the flow-invariance of D with respect to (1.4). Indeed, if $x \in D$ and F is continuous from $]a,b[\times X$ into X, then the existence of the solution is guaranteed by a theorem of Pazy [3] (see also Remark 1.3). The new aspect here is that the constraint "$u(t) \in D, \forall t \in [t_0,b_0[$" for every $t_0 \in]a,b[$ and $x \in D$ with $b_0 = b_0(t_0,x) < b$), that is the flow-invariance of D with respect to (1.30) or to (1.4), is characterized by the tangential condition (1.32), i.e., (D5).

Let us now proceed to state the result in a precise form:

Theorem 4.2 *Let* D *be a closed subset of the Banach space* X *and let* $F:]a,b[\times D \to X$ *be a continuous function which maps bounded subsets into bounded subsets. Moreover, let* A *be the infinitesimal generator of the compact* C_0 *semigroup* $S(t)$. *Then for every* $t_0 \in]a,b[$ *and* $x \in D$, *the initial value problem* (4.1) *has a noncontinuable solution* $u:[t_0,t_{max}[\to D$, *if and only if*

$$\lim_{h \downarrow 0} h^{-1} d(S(h)x + hF(t,x);D) = 0, \forall t \in]a,b[, x \in D. \tag{4.5}$$

If $t_{max} < b$ *then* $\lim_{t \uparrow t_{max}} \|u(t)\| = +\infty$.

In the sequel, the hypothesis "$S(t)$ is compact for $t > 0$" is replaced by g-dissipativity of F (see (D4) in §3). In this case, by using the proof of Theorem 2.4 in Chapter 2 (with straightforward modifications) we can establish:

Theorem 4.3 *In addition to the hypotheses of Theorem* 3.1, *suppose that* $t \to D(t)$ *is closed. Then for every* $\bar{b} \in]a,b[$ *which satisfies Hypothesis* (E2) *in* §3, *the following result holds: A necessary and sufficient condition for Equation* (1.4) *to have a solution* u *on* $[t_0,\bar{b}[$, *with* $u(t) \in D(t)$ *for all* $t \in [t_0,\bar{b}[$ *(for each* $t_0 \in]a,\bar{b}[$ *and* $x \in D(t_0)$*), is given by* (D5). *The solution* u *is unique (in view of* (D4)*).*

In particular, if $D(t) = D$ is independent of t, then (E2) is satisfied with $\bar{b} = b$ and $w = w_x(t) = x$, for all $t \in]a,b[$ and $x \in D$. Consequently, Theorem 4.3 becomes:

Theorem 4.4 *Let* D *be a closed subset of* X *and let* A *be the infinitesimal generator of the* C_0 *semigroup (as indicated by* (D1)*). Moreover, assume that* $F:]a,b[\times D \to X$ *is continuous and*

$$(1-\lambda g(t))(\|x_1-x_2\| \leq \|x_1-x_2 - \lambda(F(t,x_1)-F(t,x_2))\| \tag{4.6}$$

for all $\lambda > 0$, $t \in]a,b[$ *and* $x_1,x_2 \in D$.

Then a necessary and sufficient condition for Equation (1.4) *to have a global solution* $u:[t_0,b[\to D$ *(for every* $t_0 \in]a,b[$ *and* $x \in D$*) is given by the tangential condition* (4.5). *The solution* u *is unique (on the basis of* (4.6)*).*

Remark 4.2 Let us discuss the following two conditions:

(P) For each $t \in]a,b[$, the operator $x \to F(t,x)$ is continuous and dissipative from the closed subset $D \subset X$ into X.

(P_1) For each $t \in]a,b[$, $x \to F(t,x)$ is continuous and dissipative from X into X.

Condition (P) is strictly more general than (P_1). This is because, in general, a continuous and dissipative operator on a closed subset $D \subset X$ cannot be extended to the whole X, by preserving both continuity and dissipativity. In the case $X = R$, this is obviously possible. If X is infinite dimensional, there are partial differential operators satisfying (at least locally) (P). Alternatively, such operators cannot satisfy (P_1). For this problem we refer to Bourguignon and Brezis [1]. See also Appendix.

If Condition (P) in Theorem 4.4 is replaced by (P_1), then of course one obtains a result on the flow-invariance of D with respect to (3.1). Let us state it:

Theorem 4.5 *Let* A *be the infinitesimal generator of the* C_0 *semigroup* S(t) *as in* (D1), *and let* $F:]a,b[\times X \to X$ *be continuous and satisfying* (4.6) *with* X *in place of* D. *Then a necessary and sufficient condition for a closed subset* $D \subset X$ *to be a flow-invariant set with respect to* (4.1), *is given by the tangential condition* (4.5).

§5. Some applications to partial differential equations

Let Ω be a bounded domain in the N-dimensional Euclidean space R^N, with smooth boundary Γ (Γ is a C^k-manifold, $k \geq 1$, of N-1 dimension). A generic point $x \in R^N$ is denoted by $x = (x_1,\ldots,x_N)$. Set $D_j = \partial/\partial x_j$, $j = 1,\ldots,N$. If $\alpha = (\alpha_1,\ldots,\alpha_N)$ is an N-tuple of non-negative integers, we set

$$|\alpha| = \alpha_1 + \ldots + \alpha_N, \quad D^\alpha = \frac{\partial^{\alpha_1}}{\partial x_1^{\alpha_1}} \cdots \frac{\partial^{\alpha_N}}{\partial x_N^{\alpha_N}} = D_1^{\alpha_1} \cdots D_N^{\alpha_N}.$$

We now define the Sobolev space

$$W^{m,p}(\Omega) = \{u; D^\alpha u \in L^p(\Omega),\ |\alpha| \leq m\},\ m \geq 0,\ 1 \leq p$$

where $D^\alpha u$ is taken in the sense of distributions. In particular, $W^{o,p}(\Omega) \equiv L^p(\Omega)$. One denotes $W^{m,2}(\Omega) = H^m(\Omega)$. Recall that $W^{m,p}(\Omega)$ is a Banach space with respect to the norm

$$\|u\|_{m,p} = \Big(\sum_{|\alpha| \leq m} \int_\Omega |D^\alpha u(x)|^p dx\Big)^{1/p}.$$

$W_0^{m,p}(\Omega)$ consists of all elements of $W^{m,p}(\Omega)$ which vanish (in some generalized sense) on Γ. Set $W_0^{m,2}(\Omega) = H_0^m(\Omega)$. Let us also recall the following imbedding theorem:

Theorem 5.1 *If* j *and* m *are integers such that* $0 \leq j < m$, *and if* $1 \leq p$, $q < \infty$ *satisfy*

$$\frac{1}{p} > \frac{1}{q} + \frac{j}{N} - \frac{m}{N},$$

then $W^{m,q}(\Omega) \subset W^{j,p}(\Omega)$ *and the imbedding is compact.*

For $j = 0$, it follows that if $mq < N$ and $\frac{1}{p} > \frac{1}{q} - \frac{m}{N}$, then $W^{m,q}(\Omega) \subset L^p(\Omega)$ and the imbedding is compact (i.e., any bounded subset of $W^{m,q}(\Omega)$ is relatively compact in $L^p(\Omega)$). In particular, if $N > 2$, then $H^1(\Omega)$ is compactly imbedded in $L^2(\Omega)$.

It was shown by Brezis [2] that for the Laplace operator

$$A = \Delta,\ \Delta = D_1^2 + \ldots + D_N^2;\ D(A) = H_0^1(\Omega) \cap H^2(\Omega)$$

we have $A = -\partial\phi$, where $\partial\phi$ is the subdifferential of the proper, lower-semicontinuous, convex function (l.s.c) ϕ, given by (see (2.22) in Chapter 1):

$$\phi(u) = \begin{cases} \frac{1}{2}\int_\Omega |\text{grad } u|^2 dx, & \text{if } u \in H_0^1(\Omega) \\ +\infty, & \text{otherwise.} \end{cases} \tag{5.1}$$

A main remark of this section is given by:

Theorem 5.2 *The linear contractions semigroup* $S(t):L^2(\Omega) \to L^2(\Omega)$ *generated by* $-\Delta$ *is compact* (for $t > 0$).

To prove this assertion, we need the following theorem of Brezis [4] (which is an extension of Theorem 1.1 of Pazy):

Theorem 5.3 (Brezis). *Let* X *be a Banach space and* $A:D(A) \subset X \to X$ *be m-dissipative. The semigroup* $S(t)$ *generated by* $-A$ *via the exponential formula (i.e.,* (3.8) *in Chapter* 2) *is compact if and only if:* (1) *the resolvent* $(I-\lambda A)^{-1}$ *is compact for* $\lambda > 0$; (2) *for every bounded set* $Y \subset \overline{D(A)}$ *and* $t_0 > 0$ *the functions* $\{t \to S(t);\ x \in Y\}$ *are equicontinuous at* $t = t_0$.

If X is a Hilbert space H and $A = -\partial\phi$, this theorem yields:

Corollary 5.1 *Let* $\phi:H \to]-\infty,+\infty]$ *be* l.s.c. *The semigroup* $S(t)$ *generated by* $\partial\phi$ *is compact if and only if the level sets.*

$$C_m^\phi = \{x \in D_e(\phi);\ \|x\| \leq m,\ \phi(x) \leq m\},\ m > 0 \tag{5.2}$$

are relatively compact in H, *where* $D_e(\phi)$ *is the effective domain of* ϕ*(see*

(2.19) *in Chapter* 1).

Indeed, in this case Condition (2) of Theorem 5.3 is fulfilled and the compactness of $(I + \lambda\partial\phi)^{-1}$ is equivalent to the precompactness of C_m.

Proof of Theorem 5.2 If $H = L^2(\Omega)$ and ϕ is given by (5.1), then C_m becomes

$$C_m = \{u \in H_o^1(\Omega);\ \|u\|_2 \leq m,\ \int_\Omega |\text{grad } u|^2\, dx \leq 2m\} \qquad (5.2)'$$

which is bounded in $H^1(\Omega)$ and (consequently) precompact in $L^2(\Omega)$, by the imbedding theorem above. Here $\|u\|_2$ denotes the norm of u in $L^2(\Omega)$.

The purpose of this section is to study the problem

$$\frac{\partial u(t,x)}{\partial t} = \Delta_x u(t,x) + f(t,u(t,x)),\ \text{(a.e.) in } [0,T] \times \Omega \qquad (5.3)$$

$$u(0,x) = u_0(x) \qquad ,\ \text{(a.e.) in } \Omega \qquad (5.4)$$

$$u(t,x) = 0,\ \text{(a.e.) on } [0,T] \times \Gamma\ ,\ t^{1/2}u_t' \in L^2(0,T;L^2(\Omega)) \qquad (5.5)$$

where $u_t' = \partial u/\partial t$ and $T > 0$, while $\Delta_x = \Delta$ is the Laplace operator in R^N, as indicated above. A basic hypothesis is the continuity of f from $R_+ \times R$ into R.

We will reduce the problem (5.3) - (5.5) to the abstract problem in $L^2(\Omega)$:

$$u' = Au + F(t,u),\ u(0) = u_0,\ u_0 \in L^2(\Omega),\ t \in [0,1[\qquad (5.6)$$

where $A = \Delta$ and F is the Nemytski operator

$$(F(t,u))(x) = f(t,u(x)),\ x \in \Omega\ ,\ u \in L^2(\Omega),\ t \in [0,1]. \qquad (5.7)$$

We are of course interested in the condition

$$F:[0,1] \times L^2(\Omega) \to L^2(\Omega). \qquad (5.8)$$

Once this is accomplished, F is continuous and therefore we can apply the results of the previous sections to (5.6). Roughly speaking, by a solution to (5.3) - (5.5) we mean a solution to (5.6). In the (quite restrictive) case in which f satisfies a growth condition in x, e.g.

$$|f(t,x)| \leq c(t)\ |x|,\quad t \geq 0,\ x \in R \qquad (5.9)$$

with c a positive function, (5.8) is obviously satisfied. Accordingly, by a combination of Theorem 1.6 with Theorem 5.2 we get:

Theorem 5.4 *Let* $f: R_+ \times R \to R$ *be a continuous function which satisfies* (5.9). *Then for every* $u_0 \in L^2(\Omega)$ *Problem* (5.6) (*and therefore* (5.3) - (5.5)) *has a strong solution in* $L^2(\Omega)$ *with* $T = T(u_0) > 0$.

Proof The existence has already been proved. The property $t^{1/2} u'_t \in L^2(0,T;L^2(\Omega))$ as well as the differentiability of u (a.e. on]0,T[) is a consequence of the fact that -A is a subdifferential ($A = -\partial\phi$). In this case, $u(t) \in D(A)$ a.e. on]0,T[and satisfies (5.6) in the strong sense of $L^2(\Omega)$ (see, e.g., Barbu [2, p. 188], Brezis [3] or the author [14, p. 258]).

In what follows we drop Hypothesis (5.9) and keep only the continuity of f. Precisely, suppose that

(F1) *The function* $f: R_+ \times R \to R$ *is continuous and*

$$|f(t,y)| \leq 1, \ \forall t \in [0,1], \ |y| \leq 1.$$

In order to use Nemytski's operator F and to verify some tangential conditions, the following notation is needed:

$$B(r) = \{u \in L^2(\Omega); \|u\|_2 \leq r\}, \ B_\infty(r) = \{u \in L^2(\Omega); \|u\|_\infty \leq r\},$$

$$\overset{\circ}{B}_\infty(r) = \{u \in L^2(\Omega); \|u\|_\infty < r\}, \ r > 0,$$

where $\|\cdot\|_2$ and $\|\cdot\|_\infty$ denotes the norm of $L^2(\Omega)$ and $L^\infty(\Omega)$, respectively. If f is only continuous, then (5.8) is no longer guaranteed. Clearly, the continuity of f from $R_+ \times R$ into R guarantees

$$F: R_+ \times B_\infty(r) \to L^2(\Omega) \tag{5.10}$$

$$F: R_+ \times \overset{\circ}{B}_\infty(r) \to L^2(\Omega). \tag{5.11}$$

The situation (5.10) is not convenient for the application of Theorem 1.5 (although $B_\infty(r) \equiv D$ is closed in $L^2(\Omega)$). This is because the tangential condition (1.33) may not be satisfied in this case. In the situation (5.11) we cannot apply Theorem 1.6 in $L^2(\Omega)$, since $\overset{\circ}{B}_\infty(r)$ is not open in $L^2(\Omega)$. Moreover, $\overset{\circ}{B}_\infty(r)$ is not even locally closed in $L^2(\Omega)$. Let us demonstrate this.

Assume by contradiction that $\overset{\circ}{B}_\infty(r)$ would be locally closed in $L^2(\Omega) \equiv L^2$,

$\Omega \equiv [0,1]$. Then there would exist $d > 0$ such that $\overset{\circ}{B}_\infty(r) \cap B(d) \equiv C$ is closed in L^2. We may assume that $d < r$. Set

$$g_n(t) = \begin{cases} r - 1/n, & \text{if } 0 \leq t \leq d^2r^{-2} \\ 0, & \text{if } d^2r^{-2} < t \leq 1 \end{cases}, \quad g(t) = \begin{cases} r, & \text{if } 0 \leq t \leq d^2r^{-2} \\ 0, & \text{if } d^2r^{-2} < t \leq 1. \end{cases}$$

It is easily seen that $g_n \in C$, $g_n \to g$ (in L^2) as $n \to \infty$. However, $g \notin \overset{\circ}{B}_\infty(r)$ (since $\|g\|_\infty = r$), which contradicts the fact that C is closed in L^2. However, even in the situation (5.11) (in place of (5.8)), Problem (5.6) can be solved in $L^2(\Omega)$, as a problem with t-dependent domain, via Theorem 1.4. For this purpose we first prove:

<u>Lemma 5.1</u> *If* $f: R_+ \times R \to R$ *is continuous, then Nemytski's operator* F *given by* (5.7) *satisfies* (5.10) - (5.11) *and it is continuous. If, in addition, Hypothesis* (F1) *is fulfilled, then* $\|F(t,u)\|_\infty \leq 1$ *for all* $t \in [0,1]$, $u \in B_\infty(1)$ *and* F *satisfies the tangential condition* (1.32) *with* $a = 0$, $b = 1$, $D(t) = B_\infty(t)$ *and* u *in place of* x.

<u>Proof</u> Clearly, (5.10) and (5.11) hold. Let us show the continuity of F. We have to prove that if $u_n \to u$ in $L^2(\Omega)$ and $t_n \to t$, then $F(t_n,u_n) \to F(t,u)$. For this purpose it suffices to prove that every pair of convergent subsequences $F(t_{n_k},u_{n_k})$ and $F(t_{n_p},u_{n_p})$ of $F(t_n,u_n)$, have the same limit. Indeed, we may assume (relabelling if necessary) that $\lim_{k\to\infty} u_{n_k}(x) = \lim_{p\to\infty} u_{n_p}(x) = u(x)$, a.e. in Ω. This implies

$$\lim_{k\to\infty} (F(t_{n_k},u_{n_k}))(x) = \lim_{p\to\infty} (F(t_{n_p},u_{n_p}))(x) = f(t,u(x)), \text{ a.e. in } \Omega$$

which concludes the proof of the continuity of F.

Finally, for $t \in [0,1[$ and $u \in B_\infty(t)$ we have

$$\|S(h)u + hF(t,u)\|_\infty \leq \|u\|_\infty + h \leq t + h,$$

that is, $S(h)u + hF(t,u) \in B_\infty(t+h)$, so (1.32) is automatically satisfied. The proof is complete.

Combining Theorems 1.4 and 5.2 with Lemma 5.1, one obtains:

Theorem 5.5 *Suppose that* f *satisfies Hypothesis* (F1) *above. Then for every* $t_0 \in [0,1[$ *and* $u_0 \in B_\infty(t_0)$ *there is* $T = T(t_0,u_0)$, $0 < t < 1$ *and a continuous function* $u:[t_0,t_0+T(t_0,u_0)] \to L^2(\Omega)$ *with the properties:* $u(t) \in D(A)$, u *is differentiable, a.e. in* $]t_0,t_0+T[$ *and* (5.6) *is verified a.e. on* $]t_0,t_0+T[$ *with* $u(t_0) = u_0$. *In addition,* $t^{1/2} u_t' \in L^2(t_0,t_0+T; L^2(\Omega))$ (*i.e.,* u *is a solution to the problem* (5.3) - (5.5) *with* t_0 *and* t_0+T *in place of* 0 *and* T, *respectively*).

Remark 5.1 As we have already seen, for the existence of a mild solution u to Problem (5.6) one uses the results of Section 1. For some regularity properties of u the following theorem of Brezis has been used (see the proof of Theorem 5.4).

Theorem 5.6 *Let* H *be a Hilbert space,* $\phi:H \to]-\infty, +\infty]$ *a proper, lower-semi-continuous convex function,* $u_0 \in \overline{D(A)}$ (*where* $A = -\partial\phi$) *and* $g \in L^2(0,T;H)$. *Then there exists a unique strong solution* $u \in C([0,T];H)$ *to the initial value problem* $u'(t) = Au(t) + g(t)$, *a.e. on* $]0,T[$, $u(0) = u_0$ *with the properties:*

$$u(t) \in D(A), \text{ a.e. on }]0,T[$$
$$u \in W^{1,2}(d,T;H), \quad \forall d \in]0,T[$$
$$t^{1/2} u' \in L^2(0,T;H), \ \phi(u) \in L^1(0,T;H).$$

If in addition, $u_0 \in D_e(\phi)$, *then*

$$u' \in L^2(0,T;H) \text{ and } \phi(u) \in L^\infty(0,T).$$

We say that the semigroup S(t) generated by $\partial\phi$ has a "smoothing" effect on the initial data u_0 and g. In particular, as a consequence of this theorem, the mild solution u of (5.6) has the properties which are given in Theorem 5.6.

Remark 5.2 If, in addition to (F1), we assume that for each fixed $t \geq 0$, the function $x \to f(t,x)$ is nonincreasing, then $u \to F(t,u)$ is dissipative. Consequently, in view of Theorem 3.1, the solution given by Theorems 5.4 and 5.6 is (in this case) unique.

Consider now the following mixed problem:

$$\frac{\partial u(t,x)}{\partial t} = \frac{\partial^2 u(t,x)}{\partial x^2} + f(u(t,x)),\ t > 0,\ 0 < x < 1 \tag{5.12}$$

$$u(t,0) = u(t,1),\ u_x'(t,0) = u_x'(t,1),\ t \geq 0 \tag{5.13}$$

$$u(0,x) = u_0(x). \tag{5.14}$$

This problem describes the heat flow in a ring of length one under the action of a source f depending on the temperature u.

For the study of this problem we can also use the results in Sections 1 and 4. We will work in the space $X = X_c$, where $X_c = \{u;\ u$ continuous from $[0,1] \times R$, $u(0) = u(1)\}$ endowed with the supremum norm. Set

$$D(A) = \{u \in C^2([0,1]);\ u,u',\ u'' \in X_c\},\ Au = u'' \text{ for } u \in D(A). \tag{5.15}$$

Similarly to (5.7), one defines

$$(F(u))(x) = f(u(x)),\quad \forall u \in X_c. \tag{5.16}$$

Clearly, if $f:R \to R$ is continuous, then $F:X_c \to X_c$ is also continuous.

A few comments are necessary before we proceed to state the results on (5.12). A semigroup S(t) is said to be analytic if $z \to S(z)$ is analytic in a sector D around the non-negative real axis. Let A be the infinitesimal generator of the uniformly bounded C_0 semigroup S(t) and let $(\lambda I-A)^{-1} \equiv R(\lambda;A)$ be the resolvent of A. If there exist $d \in]0,\pi/2[$ and $K > 0$ such that $\rho(A) \supset B = \{\lambda: |\arg\lambda| \leq \frac{\pi}{2} + d\} \cup \{0\}$ and

$$||R(\lambda;A)|| \leq M/|\lambda|,\ \forall\lambda \in B,\ \lambda \neq 0 \tag{5.17}$$

then S(t) can be extended to an analytic semigroup in a sector $D_d = \{z;\ |\arg z| < d\}$. Moreover, S(t) is differentiable for $t > 0$ and there is a constant $C > 0$ such that

$$||S(t+h) - S(t)|| \leq h||AS(t)|| \leq \frac{C}{t} h,\ \forall t > 0,\ h \geq 0. \tag{5.18}$$

In particular, it follows from (5.18) that every analytic semigroup S(t) is continuous in the uniform operator topology. Finally, on the basis of Pazy's theorem (i.e., Theorem 1.1) we see that if $R(\lambda;A)$ is compact for $\lambda \in \rho(A)$ and S(t) is analytic, then S(t) is compact for $t > 0$.

We are now prepared to give:

Lemma 5.2 *The operator* A *defined by* (5.15) *is the infinitesimal generator of a compact (and analytic) semigroup* S(t) *on* X_c.

Proof Clearly $\overline{D(A)} = X_c$. Let us prove that

$$R(\lambda I - A) = X_c, \quad |\arg \lambda| < \pi/2. \tag{5.19}$$

This means that for every $g \in X_c$ we have to find $u \in D(A)$ such that

$$\begin{aligned} &\lambda^2 u - u'' = g \\ &u(0) = u(1),\ u'(0) = u'(1). \end{aligned} \tag{5.20}$$

It is easily seen that the solution $u(x) = ((\lambda^2 I-A)^{-1}g)(x) = (R(\lambda^2;A)g)(x)$ of the boundary value problem (5.19) is

$$\begin{aligned} u(x) = (2\lambda \sinh \lambda/2)^{-1} [&\int_0^x \cosh \lambda(x-y-\tfrac{1}{2})g(y)dy \\ &+ \int_x^1 \cosh \lambda(x-y+\tfrac{1}{2})g(y)dy] \end{aligned} \tag{5.21}$$

Set $\mathrm{Re}\lambda = r$ and $\arg\lambda = \theta$, hence $r = |\lambda| \cos \theta > 0$. The well-known inequalities

$$|\sinh z| \geq \sinh (\mathrm{Re}\ z),\ |\cosh z| \leq \cosh (\mathrm{Re}\ z),$$

in conjunction with an elementary integration, yield

$$\begin{aligned} |u(x)| \leq \|g\| (2|\lambda|\sinh r/2)^{-1} [&\int_0^x \cosh r(x-y-\tfrac{1}{2})dy \\ &+ \int_x^1 \cosh r(x-y+\tfrac{1}{2})dy] = \frac{\|g\|}{|\lambda|r} = \frac{\|g\|}{|\lambda|^2 \cos \theta}, \end{aligned}$$

that is,

$$\|R(\lambda^2;A)\| \leq (|\lambda|^2 \cos \theta)^{-1}, \quad |\theta| < \pi/2. \tag{5.22}$$

On the basis of (5.17), (5.22) shows that A is the infinitesimal generator of an analytic semigroup S(t). In view of (5.21) it is easy to check that $R(\lambda;A)$ is compact for $\lambda \in \rho(A)$. Indeed, let G be a bounded subset of X_c. By (5.21) we see that $(R(\lambda;A))(G)$ are equicontinuous. This is because the derivatives of the functions in $(R(\lambda;A))(G)$ are equibounded. By the Arzela criterion, $(R(\lambda;A))(G)$ is relatively compact (precompact) in X_c. In view of the comments following (5.18), $S(t)$ is compact for $t > 0$, which concludes the proof.

We turn now to (5.12). With A and F given by (5.15) and (5.16), respectively, the problem (5.12) - (5.14) reduces to the semilinear problem in X_c,

$$u'(t) = Au(t) + F(u(t)),\ u(0) = u_0 \in X_c,\ t \geq 0. \tag{5.23}$$

By Lemma 5.2 and Theorem 4.2 (with $D = X = X_c$) there follows:

<u>Theorem 5.7</u> (i) *Let* $f:R \to R$ *be a continuous function. Then for every* $u_0 \in X_c$, *the initial value problem* (5.23) *has a mild solution* $u \in C([0,t_{max}[;X_c)$, *with* $t_{max} > 0$. *If* $t_{max} < +\infty$, *then* $\lim_{t\uparrow t_{max}} \|u(t)\| = +\infty$.

(ii) *If* f *is Hölder continuous (with exponent* $r \in]0,1[$ *and* $L > 0$), *that is*

$$|f(y) - f(z)| \leq L|y - z|^r,\ y,z \in R, \tag{5.24}$$

then the mild solution u *from* (i) *is a classical solution of the problem* (5.12) - (5.14).

<u>Proof</u> (i) Obviously, the continuity of $f:R \to R$ implies that $F:X_c \to X_c$ (given by (5.16)) maps bounded subsets into bounded subsets. Thus, the first assertion of the theorem is a consequence of the compactness of $S(t)$ and of Theorem 4.2.

(ii) Set $g(t) = F(u(t))$, $0 < t < t_{max}$. If $0 < T < t_{max}$, then for $p \geq 1$, $g \in L^p(0,T;X_c)$, since g is continuous from $[0,t_{max}[$ into X_c. A well-known result (Pazy [5, p. 110]) on the regularity of a mild solution u of $u' = Au+g$ in the case where $S(t)$ is analytic, asserts that u is Hölder continuous with exponent $r_0 = (p-1)/p$, $p > 1$. On the other hand, (5.24) shows that F is also Hölder continuous with exponent r. It follows that g is Hölder continuous (with $r_1 = r(p-q)/p$, $p > 1$). This implies that the mild solution u to (5.23), with $u_0 \in D(A)$, is a strong solution (in X_c) of (5.23). It follows that the function $u(t) = u(t,x)$ is a classical solution of the problem (5.12)-(5.14) on $[0,t_{max}[$. The proof is complete.

Let us turn to the reaction-diffusion system (4.41) in Chapter 2. We will give only a result on the global existence of the solution. For this purpose, the following theorem (which is similar to Theorem 5.2) is needed:

Theorem 5.8 *Let* $j:R \to [0,+\infty]$ *be a lower-semicontinuous convex* (l.s.c.) *function. Then the semigroup* $S(t)$ *generated by the* m*-dissipative operator* $A = -\Delta$ *with*

$$D(A) = \{u \in H^2(\Omega);\ -\frac{\partial u}{\partial n} \in \partial j(u),\ a.e.\ on\ \partial\Omega\} \tag{5.25}$$

is compact for $t > 0$.

Proof By a well-known result of Brezis, $A = -\partial\psi$, where

$$\psi(u) = \begin{cases} \frac{1}{2}\int_\Omega |\text{grad } u|^2 dx + \int_{\partial\Omega} j(u)d\sigma, & \text{if } u \in H^1(\Omega) \text{ and } j(u) \in L^1(\Omega) \\ +\infty, & \text{otherwise.} \end{cases}$$

Since j has only positive values, the level set (5.2) (with ψ and $L^2(\Omega)$ in place of ϕ and H, respectively) satisfies

$$C_m^\psi \subset \{u \in H^1(\Omega);\ \|u\|_2 \leq m,\ \int_\Omega |\text{grad } u|^2 dx \leq 2m\}.$$

This shows that C_m^ψ is relatively compact in $L^2(\Omega)$ (see the proof of Theorem 5.2). On the basis of Corollary 5.1, the proof is complete.

We are now prepared to give:

Theorem 5.9 *Let* $\beta_i:R \to [0,+\infty]$, $i = 1,2$, *be* l.s.c. *and let* $g:R \to R$ *be continuous, satisfying a growth condition. Then for each* $(u_1^0,u_2^0) \in \overline{D(B)}$, *the problem* (4.41) - (4.42) *has a strong-*$L^2(\Omega)$ *solution* $u = (u_1,u_2)$ *on* $[0,+\infty[$, *having the regularity given by Theorem* 5.6 *with* $H = L^2(\Omega) \times L^2(\Omega)$.

Proof Here $B = -\text{diag}(d_1,d_2)\Delta$, with the domain

$$D(B) = \{u_1,u_2 \in H^2(\Omega);\ -\frac{\partial u_i}{\partial n} \in \partial\beta_i(u_i) \text{ on } \partial\beta,\ i = 1,2\}.$$

Set $(F(u))(x) = (-a_{11}u_1 + a_{12}u_2, g(u_1) - a_{22}u_2)$, $u_1,u_2 \in L^2(\Omega)$ with $u = (u_1,u_2)$. In view of Theorem 5.8, B generates a compact semigroup. A combination of Theorems 6.1 and 6.2 leads to the conclusion of this theorem.

§6. Notes and Remarks

The results in Section 1 are due to the author, except for Theorems 1.1 and 1.6 of Pazy. Theorems 1.2, 1.3 and 1.4 improve some results due to the author and Vrabie [3]. Theorem 1.5 was proved by the author in his paper [8]. The idea to assume the compactness of S(t) in the theory of semilinear equations (1.30) goes back to Pazy [3]. Another framework in which to obtain the existence of (1.30) is the following: let $Y \subset X$ be a subspace of X, with continuous embedding and let S(t) be a C_0-semigroup on X. Suppose that the restriction of S(t) to Y is also a C_0-semigroup on Y. We can prove the local existence of the solution to (1.30) in the space Y by assuming the compactness of S(t) only in X (i.e., in a space which contains Y). This idea is useful in applications to partial differential equations, since, for instance, it is easier to verify the compactness of S(t) in $L^2(\Omega)$ rather than in $L^\infty(\Omega)$. Some results in this direction can be found in the paper [3] by the author and I. Vrabie. Theorem 1.5 of the author was extended to the case A = A(t) by J. Prüss in his papers [1-2].

Section 2 is based on Iacob and the author [1].

The results in Section 3 are taken from the author's paper [15] and extend those in Chapter 2, Section 2. A main feature of Sections 1-3 is the technique which is employed for proving the results. So far this technique cannot be avoided. This is because our hypotheses are very general, so the existing fixed point theorems cannot be used for the proof of our results here. Under additional conditions, Theorem 4.4 has been established as follows: (1) D = X and F(t,x) = F(x) (Webb [1]); (2) $S(t): D \to D$ and F maps bounded subsets into bounded subsets (Martin [2, pp. 353-355]. More details are given in the author's paper [15]. It is interesting to point out that the conclusion of Proposition 4.3 remains valid in the case in which A is nonlinear. Let us state the result in a precise form:

Theorem 6.1 *Let* X *be a Banach space and let* $A \subset X \times X$ *be a* m-*dissipative set* (Ch. 1, §3). *Moreover, assume that:*

(1) $F: R_+ \times X \to X$ *is continuous and maps bounded sets into bounded sets.*

(2) $f \in L^1_{loc}(R_+;X)$, $R_+ = [0,+\infty[$.

(3) *For every* $t_0 \geq 0$ *and* $x_0 \in \overline{D(A)}$ *the initial value problem*

$$u'(t) \in Au(t) + F(t,u(t)) + f(t), \; u(t_0) = x_0 \tag{6.1}$$

has a local (integral) solution, that is, there exist $T = T(t_0,x_0) > 0$ *and a continuous function* $u:[t_0,t_0+T] \to \overline{D(A)}$ *such that*

$$\|u(t)-x\| \leq \|u(t_0+h)-x\| + 2\int_{t_0+h}^{t} \langle F(q,u(q))+f(q)+y,u(q)-x\rangle_+ dq \tag{6.2}$$

for all $[x,y] \in A$, $h \geq 0$ *and* $t \in [t_0+h,t_0+T]$. *Let* $u:[t_0,t_{max}[\to D(A)$ *be a noncontinuable solution. If* $t_{max} < \infty$, *then* $\lim_{t\uparrow t_{max}} \|u(t)\| = \infty$.

<u>Proof</u> We first recall the well-known inequality of Benilan [1] which gives an estimate of the difference of two integral solutions: if u_i is the integral solution of the problem

$$u_i'(t) \in Au(t) + f_i(t), \; t = t_0, \; u_i(t_0) = x_i \in \overline{D(A)}, \; i = 1,2$$

then

$$\|u_1(t) - u_2(t)\| \leq \|u_1(s) - u_2(s)\| + \int_s^t \|f_1(q) - f_2(q)\| dq, \tag{6.2)'$$

for all $0 \leq t_0 \leq s \leq t$ (see also Barbu [2], or the author [14].

The proof follows the same steps as the proof of Proposition 4.3. However, the arguments are different. Let us assume that $t_{max} < +\infty$. Then necessarily

$$\limsup_{t\uparrow t_{max}} \|u(t)\| = +\infty. \tag{6.3}$$

Indeed, if we assume that (6.3) were false, then there would exist $M_1 > 0$ such that $M_1 \geq \|F(t,u(t))\|$, $\forall t \in [t_0,t_{max}[$. Given $\varepsilon > 0$, choose $d = d(\varepsilon) > 0$ with the property

$$dM_1 + \int_{t_m-d}^{t_m} \|f(s)\| ds < \varepsilon/3, \text{ where } t_m \equiv t_{max}. \tag{6.4}$$

Denote by v the integral solution to the problem

$$v'(t) \in Av(t), \; v(t_m-d) = u(t_m-d), \; t_m-d \leq t.$$

Then, according to Benilan's inequality,

$$\|u(t) - v(t)\| \leq \int_{t_m-d}^{t_m} \|F(s,u(s))\| \, ds + \int_{t_m-d}^{t_m} \|f(s)\| \, ds$$

$$\leq dM_1 + \int_{t_m-d}^{t_m} \|f(s)\| \, ds < \varepsilon/3, \quad \forall t \in [t_m-d, t_m[. \qquad (6.5)$$

Now let $d_1 = d_1(\varepsilon) > 0$ be such that $d_1 < d$ and $\|v(t) - v(s)\| < \varepsilon/3$, for all $t, s \in [t_m-d_1, t_m]$. Then we have

$$\|u(t)-u(s)\| \leq \|u(t) - v(t)\| + \|v(t)-v(s)\| + \|v(s)-u(s)\| < \varepsilon$$

for all $t, s \in [t_m-d_1, t_m[$, that is, $\lim_{t \uparrow t_m} u(t)$ exists. This implies the contradiction that u can be extended to the right of t_{max}. Consequently, (6.3) is proved.

It remains to prove that (6.3) implies even $\liminf_{t \uparrow t_{max}} \|u(t)\| = +\infty$. Indeed, if this were false, there would exist $r > 0$ and a sequence $t_k \uparrow t_{max}$ with the properties:

$$\|u(t_k)\| \leq r, \; t_o < t_k < t_{k+1}, \; k = 1,2,\ldots \, . \qquad (6.6)$$

Let S(t) be the semigroup generated by A via the exponential formula of Crandall and Liggett (see (3.8) in Ch. 2). Since $(t,x) \to S(t)x$ maps bounded sets into bounded sets, given $b > 0$ there exists $c > 0$ such that

$$\|S(h)u(t_k)\| \leq c, \; k = 1,2,\ldots, \; h \in [0,b], \; r \leq c. \qquad (6.7)$$

Set

$$M = \sup\{\|F(t,y)\| \; ; \; t \in [0,t_{max}], \; \|y\| \leq 3c\}.$$

According to one of the hypotheses, it follows that $M > 0$. We may assume that b is sufficiently small that

$$Mb + \int_{t_m-b}^{t_m} \|f(s)\| \, ds < c. \qquad (6.8)$$

Take $t_{k_o} > t_m - b$ for a positive integer k_o. Then we have

$$\|u(t)\| < 3c, \; \forall t \in [t_{k_o}, t_m[. \qquad (6.9)$$

The proof of (6.9) is also by contradiction. Therefore, if (6.9) were false, there would exist $\bar{h} > 0$ satisfying

$$\bar{h} \leqslant b, \ \|u(t_{k_0}+s)\| < 3c, \ \forall s \in [0,\bar{h}[, \ t_{k_0} + \bar{h} < t_m, \ \|u(t_{k_0}+\bar{h})\| = 3c. \tag{6.10}$$

Let v be the integral solution to the problem

$$v'(t) \in Av(t), \ v(t_{k_0}) = u(t_{k_0}), \ t \geqslant t_{k_0},$$

that is, $v(t) = S(t-t_{k_0})u(t_{k_0})$, $t \geqslant t_{k_0}$. Benilan's inequality yields

$$\|u(t_{k_0}+\bar{h}) - v(t_{k_0}+\bar{h})\| \leqslant \int_{t_{k_0}}^{t_{k_0}+\bar{h}} \|F(s,u(s)) + f(s)\| \, ds < c \tag{6.11}$$

where (6.8) has also been used. Combining (6.7) with (6.11), one obtains the contradiction

$$3c = \|u(t_{k_0}+\bar{h})\| \leqslant \|u(t_{k_0}+\bar{h}) - v(t_{k_0}+\bar{h})\| + \|v(t_{k_0}+\bar{h}) - v(t_{k_0})\| < 3c,$$

which proves (6.9). On the other hand, (6.9) contradicts the unboundedness of u on $[t_0,t_{max}[$ and therefore $\liminf_{t\uparrow t_{max}} \|u(t)\| = \infty$. The proof is complete.

Except for minor modifications, the proof of Theorem 6.1 is taken from the author's paper [18]. There we have also supposed the additional hypothesis that $t \to S(t)x$ is equicontinuous with respect to x in bounded subsets of $\overline{D(A)}$, at a point $t_0 > 0$. This additional hypothesis was not essentially used in the above paper and, according to Theorem 6.1, it can be dropped. This remark is due to Vrabie (private communication).

As far as the local existence of the integral solution to (6.1) is concerned, we recall the following nonlinear version of Pazy's local existence theorem (see Theorem 1.6):

Theorem 6.2 (Vrabie [1]). *Suppose that the following conditions are fulfilled:* (1) *The semigroup* S(t) *generated by* A *via the exponential formula of Crandall and Liggett, is compact for* $t > 0$.
(2) $F:R_+ \times X \to X$ *is continuous, while* $f \in L^1_{loc}(R_+;X)$. *Then Problem* (6.1) *has a local (integral) solution.*

Theorems 4.2 - 4.5 are established by the author [11] Theorem 5.7 is due to Pazy [5, p. 235]. Finally, Theorems 5.4 - 5.5 are proved by the author. The differential inclusions (Equations (1.5) and (1.6)) arise in control theory, as closed-loop systems that the optimal arc must satisfy (Pavel & Vrabie [3]). A detailed study of differential inclusions can be found in the recent book by Aubin & Cellina [1]. Further references are Attouch and Damlamian [1], Castaing & Valadier [1], Lightbourne & Martin [1], Schiaffino [1], Zaidman [1] and others.

We now show that the technique of the proof of Theorem 6.1 in Chapter 2 can also be used to prove some basic results on the existence of (integral) solutions to the problem $u' \in Au + Bu$. For the sake of self-containment, we first prove:

<u>Lemma 6.1</u> *Let* $f:X \to X$ *be a locally Lipschitz function and let* $x_0 \in X$, $r > 0$ *and* $M > 0$ *be such that*

$$\|f(x)\| \leq M, \quad \forall \|x - x_0\| \leq r.$$

Suppose that $T > 0$ *satisfies*

$$MT + \sup\{\|S(t)x_0 - x_0\| \, , \, t \in [0,T]\} < r \tag{6.12}$$

where $S(t)$ *is the semigroup generated by the* m-*dissipative set* $A \subset X \times X$. *Then the problem*

$$u' \in Au + f(u), \; u(0) = x_0, \; x_0 \in \overline{D(A)} \tag{6.13}$$

has a unique solution $u:[0,T] \to \overline{D(A)}$ (*i.e.,* u *is continuous and satisfies*)

$$\|u(t) - x\| \leq \|u(h)-x\| + \int_h^t \langle f(u(s)) + y, u(s)-x\rangle_+ ds, \tag{6.13)'$$

for all $[x,y] \in A$, $0 \leq h \leq t \leq T$ *and* $u(0) = x_0$.

<u>Proof</u> Let $r_0 \leq r$ be such that f is Lipschitz on the closed ball $B(x_0,r_0)$ (of Lipschitz constant L) and let T_0 satisfy (6.12) with r_0 in place of r, as well as $T_0 < L^{-1}$. Set

$$K = \{v \in C([0,T];X); \; \|v(t)-x_0\| \leq r_0, \quad \forall t \in [0,T_0]\}.$$

In view of the Benilan theorem, for each $v \in K$ there exists a unique integral solution Qv of the problem

$$(Qv)' \in A(Qv) + f(v),\ (Qv)(0) = x_0,\ t > 0.$$

By the Benilan inequality (6.2)', we have

$$\|(Qv)(t) - S(t)x_0\| \leq \int_0^t \|f(v(s))\|\, ds \leq MT_0,\ t \in [0,T_0]. \tag{6.14}$$

Consequently, $Q:K \to K$. Moreover, Q is a contraction. Indeed,

$$\|(Qv_1)(t) - (Qv_2)(t)\| \leq \int_0^t \|f(v_1(s)) - f(v_2(s))\|\, ds,\ t \in [0,T_0]$$

for all $v_1, v_2 \in K$. Let $u \in K$ be the unique fixed point of Q. Since $Q:K \to K \cap \overline{D(A)}$, the function $u = Qu$ is a solution of (6.13) on $[0,T_0]$. The uniqueness also follows from Benilan inequality.

Since $r_0 \leq r$, we have $T_0 \leq T$. Moreover, since $u = Qu$ satisfies (6.14) with u in place of v and Qv, it follows that

$$\|u(t) - x_0\| \leq MT + \|S(t)x_0 - x_0\| < r,\ \ \forall t \in [0,T_0]. \tag{6.15}$$

Now let t_m be the supremum of all such T_0 and let u_m be the solution to (6.13) on $[0,t_m[$. Since u_m satisfies (6.15), we have $\|f(u_m(t))\| \leq M$ for all $t \in [0,t_m[$. Let $\varepsilon \in]0,t_m[$. Set $t_0 = t_m - \varepsilon$ and $w(t) = S(t-t_0)u_m(t_0)$. Therefore

$$\|u_m(t) - w(t)\| \leq \int_{t_0}^t \|f(u_m(s))\|\, ds \leq M(t-t_0) = M\varepsilon,\ \forall t \in [t_0,t_m[$$

which yields

$$\|u_m(t) - u_m(s)\| = 2M\varepsilon + \|w(t) - w(s)\|,\ \ \forall t,s \in [t_m-\varepsilon,t_m[,$$

that is, $\limsup_{t,s\uparrow t_m} \|u_m(t) - u_m(s)\| \leq 2M\varepsilon$, hence $\lim_{t\uparrow t_m} u_m(t) = x^*$ exists. It follows that $t_m = T$ and that u_m is the solution to (6.13) on $[0,T]$, q.e.d.

We are now prepared to prove:

<u>Theorem 6.3</u> (Barbu [1]). *Assume that* $B:X \to X$ *is continuous and dissipative and* $A \subset X \times X$ *is* m-*dissipative. Then for every* $x_0 \in \overline{D(A)}$ *the problem*

$$u' \in Au + Bu,\ u(0) = x_0,\ t > 0 \tag{6.16}$$

has a unique integral solution.

Proof Let $B_n: X \to X$ be locally Lipschitz as in the proof of Theorem 6.1, Chapter 2 and let T be as in (6.12). On the basis of the previous lemma, there exists a unique integral solution $u_n: [0,T] \to \overline{D(A)}$ of the problem

$$u_n' \in Au_n + B_n u_n, \; u_n(0) = x_0, \; t \in [0,T] \tag{6.17}$$

$$\|u_n(t) - x_0\| < r, \; t \in [0,T].$$

Using

$$\|u_n(t) - u_m(t)\| \leq \int_0^t \langle B_n u_n(s) - B_m u_m(s), u_n(s) - u_m(s)\rangle_+ ds$$

and arguing as in Chapter 2, §6, we conclude that the integral solutions $\{u_n\}$ satisfy Estimate (6.4) from Chapter 2. Then $u(t) = \lim_{n\to\infty} u_n(t)$ is the unique (integral) solution on [0,T] of (6.16). The extendability of u from [0,T] to $[0,+\infty[$ follows standard arguments. For a similar proof, see Pierre [1]. With the same technique we prove the following particular case of Theorem 6.2, that is:

Theorem 6.4 *If* $B: X \to X$ *is continuous and* S(t) (*generated by the* m-*dissipative set* $A \subset X \times X$) *is compact for* $t > 0$, *then* (6.16) *has at least a solution on* [0,T] (*with* T *as in* (6.12)).

Proof The only fact we have to check is that in this case the sequence of integral solutions $\{u_n\}$ is relatively compact in $C = C([0,T];X)$. Let $t \in]0,T]$ and $0 < d < t$. Set $v_n(s) = S(s-(t-d))u_n(t-d)$, for $t-d \leq s \leq T$. Hence $v_n(t-d) = u_n(t-d)$ and $v_n(t) = S(d)u_n(t-d)$. Since S(d) is compact and $\|u_n(t-d) - x_0\| < r$ for all $n = 1,\ldots$, it follows that $\{v_n(t), n = 1,\ldots\}$ is relatively compact in X. On the other hand,

$$\|u_n(s) - v_n(s)\| \leq \int_{t-d}^{s} \|B_n u_n(q)\| \, dq, \quad \forall s \in [t-d, t+d]. \tag{6.18}$$

This yields $\|u_n(t) - v_n(t)\| \leq dM$, so $\{u_n(t), n = 1,\ldots\}$ is also relatively compact. Moreover, if $s \in [t-d, t+d] \subset [0,T]$, then a standard device in conjunction with (6.18) leads to

$$\|u_n(t) - u_n(s)\| \leqslant 3dM + \|S(s-t+d)u_n(t-d) - S(d)u_n(t-d)\| \tag{6.19}$$

for $t \leqslant s \leqslant t+d \leqslant T$. The compactness of $S(d)$ implies that $t \to S(t)x$ is continuous at $t = d$, uniformly with respect to x in bounded subsets of $\overline{D(A)}$. Accordingly, (6.19) shows that equicontinuity of $\{u_n\}$. Therefore $\{u_n\}$ is relatively compact in C, q.e.d.

Appendix

The theory of differential equations on closed subsets leads to the following problem (see author's paper [13]).

(P) Let D be a nonempty closed subset of the Banach space X and let $A:D \to X$ be a continuous and dissipative function, which is tangent to D (see (3.9), Ch. 2). Can A be extended to X preserving both continuity and dissipativity?

The answer to this question is given by

<u>Theorem 1</u> *Let* $A:D \to X$ *be as in* (P). *If* X = R (real axis) *then* A *can be extended to* X *preserving both continuity and dissipativity. This fact may not be true if the dimension of* X *is greater than one.*

<u>Proof</u> Let X = R. In this case the fact that $A:D \subset R \to R$ is dissipative means that A is nonincreasing. Denote by D_1 the complementary $R \setminus D$ of D. Take an arbitrary open interval $]a,b[\subset D_1$ with $-\infty < a < b < +\infty$. Hence $a,b \in D$ and $Aa \geqslant Ab$. Set

$$Ax = Aa + (b-a)^{-1}(Ab-Aa)(x-a), \quad x \in [a,b]. \tag{1}$$

In this way, A can be extended to R as a continuous and nonincreasing function, q.e.d.

If the dimension of X is greater than one (i.e., $\dim X \geqslant 2$), then we have the following counterexample to the extendability of A to X, namely:

$$X = R^2, \quad D = \{(x,y);\ x \in R,\ y \geqslant 1/2\}, \quad A(x,y) = \left(-\frac{2x}{y}, \frac{x^2}{y^2}\right) \tag{2}$$

Clearly, -A is the gradient of the convex function

$$f(x,y) = \frac{x^2}{y}, \quad x \in R,\ y > 0 \tag{3}$$

and therefore A is dissipative. Moreover, A is tangent to D. Indeed, we have $(x,1/2) + hA(x,1/2) \in D$ for all $x \in R$ and $h > 0$, hence (3.9)' in Chapter 2 is verified. We now prove that A cannot be extended to X as a dissipative function. For this purpose, assume by contradiction that there exists a dissipative extension (denoted also by A) of A to the whole X. Set $A(0,-1) = (p,q)$ with $p,q \in R$. Then the dissipativity of A yields

$$\langle A(x,y) - A(0,-1),(x,y+1)\rangle \leqslant 0, \quad \forall x \in R,\ y \geqslant 1/2 \tag{4}$$

where $\langle\cdot,\cdot\rangle$ denotes the inner product of R^2. In view of (2), the inequality (4) becomes

$$\frac{x^2}{y}\left(\frac{1}{y} - 1\right) + px + q(y+1) \leqslant 0,\ \forall x \in R,\ y \geqslant 1/2. \tag{5}$$

Substituting $y = 1/2$ into (5), we reach the absurdity

$$4x^2 + 2px + 3q \leqslant 0, \quad \forall x \in R \tag{6}$$

which completes the proof.

Remark 1 We have actually proved that grad f cannot be extended to R^2. It follows that if $F:R^2 \to R^2$ is a function of class C^1, such that its restriction F/D to a closed convex subset $D \subset R^2$ is convex, then F/D cannot be extended to R^2 as a convex function of class C^1. This remark is due to Barbu. We have to recall that even if X = R, a continuous and convex function $F:D \subset R \to R$ (with D closed and convex) cannot be extended to R preserving both continuity and convexity. A standard counterexample in this direction is the following: $D = [-1,1]$, $F(x) = -(1-x^2)^{1/2}$, $x \in D$.

Remark 2 Obviously, A given by (2) is continuous and dissipative on the open subsets $U = \{(x,y);\ y > 0\} \supset D$. Consequently, the following problem arises:

(P_1) Let D and A be as in (P) and let $\dim X \geqslant 2$. Is it true that A cannot be extended to any open subset $U \supset D$ preserving both continuity and dissipativity?

If the answer to the above question is affirmative (as it seems to be), then use of the sequence v_ε in the proof of Theorem 3.1, Ch. 2, cannot be avoided (see Remark 3.3). Consequently, (P_1) is an interesting *open* problem.

References

Abraham, R. and Marsden, J.E.

[1] *Foundation of Mechanics.* Benjamin Cummings)1978).

Asplund, E.

[1] Averaged norms. *Israel J. Math.* 5 (1967) 227-233.

Attouch, H. and Damlamian, A.

[1] On multivalued evolution equations in Hilbert spaces. *Israel J. Math.* 12 (1972) 373-390.

Aubin, J.P. and Cellina, A.

[1] *Differential Inclusions.* Springer-Verlag (1984).

Barbu, V.

[1] Continuous perturbations of non-linear m-accretive operators in Banach spaces. *Boll. Un. Mat. Ital.* 6 (1972) 270-278.

[2] *Nonlinear Semigroups and Differential Equations in Banach Spaces.* Noordhoff (1976).

Benilan, Ph.

[1] Equations d'évolution dans un espace de Banach quelconque et applications. (Thèse, Orsay 1972).

[2] Solutions intégrales d'équations d'évolution dans un espace de Banach. *C.R. Acad. Sci. Paris Ser. A* 274 (1972) 47-50.

Bielecki, A.

[1] Un remark sur la méthode de Banach-Cacciopoli-Tikhonov. *Bull. Acad. Polon. Sci. Math.* 4 (1956) 261-264.

Bouligand, G.

[1] Sur les surfaces dépourvues de points hyperlimités. *Ann. Soc. Polon. Math.* 9 (1930) 32-41.

Bourbaki, N.

[1] *Variétés Différentiables et Analytiques (Fascicule de resultats).* Hermann, Paris (1967; 1971).

Bourceanu, G and Morosanu, G.

[1] Sisteme diferentiale care descriu procese chimice (to appear).

Bourguignon, J.P. and Brezis, H.

[1] Remarks on the Euler equation. *J. Funct. Anal.* 15 (1974) 341-363.

Brezis, H.

[1] On a characterization of flow-invariant sets. *Comm. Pure Appl. Math.* 23 (1970) 261-263.

[2] Problèmes unilatéraux. *J. Math. Pures Appl.* 51 (1972) 1-164.

[3] *Opérateurs Maximaux Monotones et Semigroupes de Contractions dans les Espaces de Hilbert.* Math. Studies 5, North Holland (1973).

[4] New results concerning monotone operators and nonlinear semigroups. *Proc.* RIMS *"Analysis of Nonlinear Problems"* Kyoto Univ. (1975).

Brezis, H. and Browder, F.E.

[1] General ordering principle in nonlinear functional analysis. *Advances in Math.* 21 (1976) 355-364.

Browder, F.E.

[1] Nonlinear equations of evolution. *Ann. Math.* 80 (1964) 485-523.

[2] *Nonlinear Operators and Nonlinear Equations of Evolution in Banach Spaces.* Proc. Sympos. Pure Math. 18, part 2, (1968), AMS Providence.

Calvert, B.

[1] Maximal accretive is not m-accretive. *Boll. Un. Mat. Ital.* 6 (1970) 1042-1044.

Capasso, V. and Paveri-Fontana, S.L.

[1] A mathematical model for the 1973 cholera epidemic in the European Mediterranean region. *Rev. Epid. et Santé Publ.* 27 (1979) 121-132. *Ibidem* 28 (1980) 390.

Capasso, V. and Maddalena, L.

[1] Convergence to equilibrium states for a reaction-diffusion system modelling the spatial spread of a class of bacterial and viral diseases. *J. Math. Biology* 13 (1981) 173-184.

Cartan, H.

[1] *Calcul Différentiel.* Paris (1967).

Castaing, C. and Valadier, M.

[1] *Convex Analysis and Measurable Multifunctions.* Lecture Notes in Math. 580 (1977), Springer-Verlag.

Cellina, A.

[1] On the non-existence of solutions of ordinary differential equations in nonreflexive spaces. *Bull. Amer. Math. Soc.* 78 (1978) 1069-1070.

[2] Differential equations in Banach spaces. Autumn Math. Course, Trieste (Italy, 1974).

Cernes, A.

[1] Ensembles maximaux accrétifs et m-accrétifs. *Israel J. Math.* 19 (1974) 335-348.

Clarke, F.H.

[1] Generalized gradients and applications. *Trans. Amer. Math. Soc.* 205 (1975) 247-262.

[2] The Euler-Lagrange differential inclusions. *J.Differential Equations* 19 (1975) 80-90.

[3] The maximum principle under minimal hypotheses. *Siam. J. Control Optimiz.* 14 (1976) 1078-1091.

[4] A new approach to Lagrange multipliers. *Math.Operations Res.* 1 (1976) 165-174.

Clarkson, J.A.

[1] Uniformly convex spaces. *Trans. Amer. Math. Soc.* 40 (1936) 396-414.

Conti, R. and Sansone, G.

[1] *Nonlinear Differential Equations.* Pergamon Press, Oxford (1964).

Crandall, M.G.

[1] A generalization of Peano's existence theorem and flow-invariance. *Proc. Amer. Math. Soc.* 36 (1972) 151-155.

[2] A generalized domain for semigroup generators. *Proc. Amer. Math. Soc.* 37 (1973) 434-440.

Crandall, M.G. and Liggett, T.M.

[1] Generation of semi-groups of nonlinear transformations on general Banach spaces. *Amer. J. Math.* 93 (1971) 265-298.

Dainelli, U.

[1] Sul movimento per una linea qualqunque. *Giornale di Matematische.* 18 (1880) 271-300.

Da Prato, G.

[1] *Applications Croissantes et Equations d'Evolution dans les Espaces de Banach.* Istituto Naz. di Alta Mat. Vol. II (1976).

Day, M.M.

[1] On the basis problem in normed spaces. *Proc. Amer. Math. Soc.* 13 (1962) 655-658.

De Blasi, F.S. and Myjak, J.

[1] La convergence des approximations successives pour les équations differentielles dans les espaces de Banach est une propriété générique. *C.R. Acad. Sci. Paris* 286 (1978) 29-31.

Deimling, K.

[1] *Ordinary Differential Equations in Banach Spaces*. Lecture Notes in Math. Springer-Verlag 596 (1977).

Dieudonne, J.

[1] Deux examples d'équations différentielles. *Acta Sci. Math. (Szeged.)* 12B (1950) 38-40.

[2] *Foundation of Modern Analysis*. Academic Press, New York (1960).

Dugundji, J.

[1] An extension of Tietze's theorem. *Pacific J. Math.* 1 (1951) 353-367.

Engler, H.

[1] Invariant sets for functional differential equations in Banach spaces and applications. Preprint 83 (1970) Univ. Heidelberg.

Federer, H.

[1] *Geometric Measure Theory*. Springer-Verlag (1969).

Field, R.J. and Noyes, R.M.

[1] Oscillations in chemical systems. IV. Limit cycle behaviour in a model of a real chemical reaction. *J. Chem. Phys.* 60 (1974) 1877-1884.

Fitzgibbon, W.E.

[1] Semilinear functional differential equations *J. Diff. Equations*. 29 (1978) 1-14.

Flett, T.M.

[1] Some applications of Zygmund's lemma to nonlinear differential equations in Banach and Hilbert spaces. *Studia Math.* XLIV (1972) 335-344.

Gheorghiev, Gh. and Oproiu, V.

[1] *Geometrie Diferentiala*. Bucuresti (1974).

Gherasim, O.

[1] A FORTRAN program for (CWPAS) of homogeneous linear constant differential systems (to appear).

Godunov, A.N.

[1] Counterexample to Peano's theorem in infinite dimensional Hilbert spaces (Russian). *Vestnik Mosk. Gosud. Univ.* 5 (1972) 31-34.

[2] On Peano's theorem in Banach spaces (Russian). *Functional. Anal. i Prilozen.* 9 (1975) 59-60.

Goldstein, J.A.

[1] Abstract evolution equations. *Trans. Amer. Math. Soc.* 14 (1969) 159-186.

Hale, J.K.

[1] *Functional Differential Equations.* Appl. Math. Sci. 3 (1971), Springer-Verlag.

Hanusse, P.

[1] *C.R. Acad. Sci. Paris, Ser. C.* 277 (1973) 263-266.

Haraux, A.

[1] *Nonlinear Evolution Equations. Global Behaviour of Solutions.* Lecture Notes in Math. 841 (1981), Springer-Verlag.

Hartman, P.

[1] On invariant sets and on a theorem of Ważewski. *Proc. Amer. Math. Soc.* 32 (1972) 511-520.

Hastings, S.P. and Murray, J.D.

[1] The existence of oscillatory solutions in the Field-Noyes model for the Belousov-Zhabotinskii reaction. *Siam. J. Appl. Math.* 28 (1975) 678-688.

Henry, D.

[1] *Geometric Theory of Semilinear Parabolic Equations.* Lecture Notes in Math. 840 (1981), Springer-Verlag.

Iacob, F. and Pavel, N.H.

[1] Invariant sets for a class of perturbed differential equations of retarded type. *Israel J. Math.* 28 (1977) 254-264.

Kachurovskii, R.I.

[1] On monotone operators and convex functionals. *Uspehi Mat. Nauk.* 15 (1960) 213-215.

Kartsatos, A.G

[2] Perturbations of m-accretive operators and quasi-linear evolution equations. *J. Math. Soc. Japan* 30 (1978) 75-84.

Kato, S.

[1] On the global existence of unique solutions of differential equations in a Banach space. *Hokkaido Math. J.* 7 (1978) 58-73.

Kato, T.

[2] Nonlinear semigroups and evolution equations. *J. Math. Soc. Japan* 19 (1967) 508-520.

Kenmochi, N. and Takahashi, T.

[1] On the global existence of solutions of differential equations on closed subsets of a Banach space. *Proc. Japan Acad. Ser. A Math. Sci.* 51 (1975) 520-525.

[2] Nonautonomous differential equations in Banach spaces. *Nonlinear Anal.* 4 (1980) 1109-1121.

Kirk, W.A.

[1] A fixed point theorem for mappings which do not increase distances. *Amer. Math. Monthly.* 72 (1965) 1004-1006.

Kobayashi, Y.

[1] Difference approximation of Cauchy problems for quasi-dissipative operators and generation of nonlinear semigroups. *J. Math. Soc. Japan.* 27 (1975) 641-663.

Lang, S.

[1] *Introduction to Differentiable Manifolds.* Interscience, New York (1962).

Lakshmikantham, V. and Leela, S.

[1] *Nonlinear Differential Equations in Abstract Spaces.* Pergamon Press, Oxford-New York (1981).

Larrieu, M.

[1] Invariance d'une ferme pour un champ de vecteurs de Carathéodory. *Publ. Mathématiques de PAU* (1981).

Lasota, A. and Yorke, J.A.

[1] The generic property of existence of solutions of differential equations in Banach spaces. *J. Diff. Equations.* 13 (1973) 1-13.

Lightbourne, J.H. and Martin, R.H. Jr.

[1] Relatively continuous nonlinear perturbations of analytic semigroups. *Nonlinear Anal.* 1 (1977) 277-292.

Lovelady, D.L. and Martin, R.H. Jr.

[1] A global existence theorem for a nonautonomous differential equation in a Banach space. *Proc. Amer. Math. Soc.* 35 (1972) 445-449.

Marshall, Ch.D.

[1] Calculus on subcartesian spaces. *J. Differential Geometry* 10 (1975) 551-578.

Martin, R.H. Jr.

[1] A global existence theorem for autonomous differential equations in a Banach space. *Proc. Amer. Math. Soc.* 26 (1970) 307-314.

[2] Differential equations on closed subsets of a Banach space. *Trans. Amer Math. Soc.* 179 (1973) 399-414.

[3] *Nonlinear Operators and Differential Equations in Banach Spaces.* Wiley, New York (1976).

McCuskey, S.W.
[1] Introduction to Celestial Mechanics. Addison-Wesley (1963).
Minty, G.J.
[1] Monotone nonlinear operators in Hilbert spaces. *Duke Math. J.* 29 (1962) 341-346.
Motreanu, D. and Pavel, N.H.
[1] Quasi-tangent vectors in flow-invariance and optimization problems on Banach manifolds. *J. Math. Anal. Appl.* 88 (1982) 116-132.
[2] Flow-invariance for second order differential equations on Banach manifolds and orbital motions. *Preprint Series in Math.* 1 (1983), Univ. Iasi.
[3] Flot-invariance par rapport aux équations différentielles du second ordre sur variété. *C.R. Acad. Sc. Paris.* 297 (1983) 157-160.
Murray, J.D.
[1] On a model for the temporal oscillations in the Belousov-Zhabotinskii reaction. *J. Chem. Phys.* (1974).
[2] *Lectures on Nonlinear Differential Equations Models in Biology.* Clarendon Press, Oxford (1977).
Nagumo, M.
[1] Über die Lage der Integralkurven gewöhnlicher Differentialgleichungen. *Proc. Phys. Math. Soc. Japan* 24 (1942) 551-559.
Palais, R.
[1] *Lectures on the Differential Topology of Infinite Dimensional Manifolds.* Brandeis Univ. Mimeographed Notes. (1965).
Pars , L.A.
[1] *Introduction to Dynamics.* Cambridge 1953.
Pavel, N.H.
[1] Sur certaines équations différentielles abstraites. *Boll. Un. Mat. Ital.* 6 (1972) 397-409.
[2] Equations non-linéaires d'évolution. *Mathematica Cluj.* 14 (1972) 289-300.
[3] On an integral equation. *Rev. Roumaine Math. Pures Appl.* 19 (1974) 237-244.
[4] Approximate solutions of (CP) for some differential equations on Banach spaces. *Funkcial. Ekvak.* 17 (1974) 85-94.

[5] *Nonlinear Evolution Equations Associated with Accretive Operators in General Banach spaces*. Mimeographed Notes, Univ. Iasi (1974).

[6] Second order differential equations on closed sets of a Banach space. *Boll. Un. Mat. Ital.* 12 (1975) 348-353.

[7] Integral solutions for nonlinear evolution equations on Banach spaces. *Studia Mathematica*. 55 (1976) 141-149.

[8] Invariant sets for a class of semilinear equations of evolution. *Nonlinear Anal.* 1 (1977) 187-196.

[9] *Ecuaţii Diferenţiale Asociate Unor Operatori Neliniari pe Spaţii Banach*. Edit. Acad. Bucharest (1977).

[10] Nonlinear evolution equations governed by f-quasi-dissipative operators. *Nonlinear Anal.* 5 (1981) 449-468; *Ibid.* 5 (1981) 1389.

[11] Global existence for nonautonomous perturbed differential equations. Preprint 22 (1981), INCREST-Bucharest.

[12] Toward the unification of the theory of nonlinear semigroups. *Bul. Inst. Politehnic Iasi*. 27 (1981) 35-40.

[13] Differential equations on closed subsets of a Banach space and applications (in *Evolution Equations and their Applications*, Kappel and Schappacher editors, Pitman Research Notes in Math. 68 (1982) 146-165.

[14] *Analysis of Some Nonlinear Problems in Banach Spaces and Applications*. Mimeographed Notes, Univ. Iasi (1982).

[15] Semilinear equations with dissipative t-dependent domain perturbations. *Isreal J. Math.* 46 (1983) 103-122.

[16] *Positivity and Stability of Some Differential Systems from Biomathematics*. Internal Report 101 (1983), ICTP, Trieste, Italy.

[17] Invariant boxes and stability of some dynamical systems from Biomath. and chemical reactions (to appear).

[18] *Some Problems on Nonlinear Semigroups and the Blow-up of Integral Solutions*. Internal Report 70 (1983), ICTP, Trieste, Italy.

Pavel, N.H. and Turinici, M.

[1] Positive solutions of a system of ordinary differential equations. *An. Stiint. Univ. Iasi*. 24 (1978) 63-70.

Pavel, N.H. and Ursescu, C.

[1] Existence and uniqueness for some nonlinear functional equations in a Banach space. *An. Stiint. Univ. Iasi* 20 (1974) 53-58.

[2] Flow-invariance for higher order differential equations. *An. Stiint. Univ. Iasi*. 24 (1978) 91-100.

[3] Flow-invariant sets for autonomous second order differential equations and applications in Mechanics. *Nonlinear Anal.* 6 (1982) 35-74.

Pavel, N.H. and I.I. Vrabie

[1] Equations d'évolution multivoques dans des espaces de Banach. *C.R. Acad. Sci. Paris* 287 (1978) 315-317.

[2] Differential equations with time-dependent domain (*Proc. Helsinki Symp. on Volterra Eq.*) Lecture Notes in Math. Londen and Staffans edit. 737 (1979) 236-249, Springer-Verlag.

[3] Semilinear evolution equations with multivalued right hand side in Banach spaces. *An. Stiint. Univ. Iasi* 25 (1979) 137-157.

[4] Flow-invariance for differential equations. Proc. Romanian-American Seminar (Mayer & Pavel org.); *An. Stiint. Univ. Iasi, Supl.* 25 (1979) 126-132.

Pascali, D. and Sburlan, S.

[1] *Nonlinear Mappings of Monotone Type.* Sijthoff & Noordhoff (1978).

Pazy, A.

[1] On the differentiability and compactness of semigroups of linear operators. *J. Math. Mech.* 11 (1968) 1131-1142.

[2] Asymptotic behaviour of the solution of an abstract evolution equation and some applications. *J. Diff. Eqns.* 4 (1968) 493-509.

[3] A class of semi-linear equations of evolution. *Israel J. Math.* 20 (1975) 23-36.

[4] On the Lyapunov method for evolution equations governed by accretive operators (in *Equations and their Applications*. Kappel & Schappacher edit), Pitman Research Notes in Math. 68 (1982) 166-189.

[5] *Semigroups of Linear Operators and Applications to Partial Differential Equations.* Springer-Verlag (Applied Math. Sci.) 44 (1983).

Pianigiani, G.

[1] A density result for differential equations in Banach spaces. *Bull. Acad. Polon. Sci.* 26 (1978) 791-793.

Pierre, M.

[1] *Génération et Perturbation de Semi-groupes de Contractions Non Lineaires.* Thèse, Univ. Paris VI (1976).

Prüss, J.

[1] *Semilineare Integrodifferentialgleichungen.* Thesis, Paderborn (1978).

[2] On semilinear parabolic evolution equations on closed sets. *J. Math. Anal. Appl.* 77 (1980) 513-538.

Redheffer, R.M.

[1] The theorems of Bony and Brezis on flow-invariant sets. *Amer. Math. Monthly* 79 (1972) 740-746.

Rockafellar, R.T.

[1] *Convex Analysis*. Princeton: Univ. Press (1970).

Roger, F.

[1] Les propriétés tangentielles des ensembles euclidiens de points. *Acta Math*. 69 (1938) 99-133.

Rubinov, S.I.

[1] *Introduction to Mathematical Biology*. New York Interscience (1975).

Schiaffino, A.

[1] Compactness methods for a class of semi-linear equations of evolution. *Nonlinear Anal*. 2 (1978) 179-188.

Scorza Dragoni, G.

[1] Una applicazione della quasicontinuita semiregolare delle funzioni misurabili rispetto ad una e continue rispetto ad un'altra variabile. *Atti. Acad. Naz. Lincei Rend. Sci. Fis. Mat. Natur*. 12 (1952) 55-61.

Severi, F.

[1] Su alcune questioni di topologia infinitesimale. *Ann. Polon. Soc. Math*. 9 (1930) 97-108.

Ursescu, C.

[1] Tangent sets and differentiable functions. *Banach Center Publications* 1 (1974) 151-155.

[2] Caratheodory solutions of ODE on locally compact sets in Frechet spaces. (Preprint 18 (1982), Univ. Iasi).

Vainberg, M.M.

[1] On the convergence of the process of steepest descent for nonlinear equations. *Sibirsk. Math. J*. 2 (1961) 201-220.

Vidossich, G.

[1] *Existence, Comparison and Asymptotic Behaviour of Solutions of Ordinary Differential Equations in Infinite Dimensional Banach Spaces*. Notas de Matematica 24, Univ. Brazilia (1972).

[2] Global convergence of successive approximations. *J. Math. Anal. Appl*. 45 (1974) 285-292.

Voicu, M.

[1] Component-wise asymptotic stability of the linear constant dynamical systems. (Preprint 5 (1982), Univ. Iasi.

[2] On the determination of the linear state feedback matrix. *Proc. Internat. Conf. on Control & Computer Science, Bucharest* (1983) Vol. I, 119-123.

Vrabie, I.I.

[1] The nonlinear version of Pazy's local existence theorem. *Israel J. Math.* 32 (1979) 221-235.

[2] Compactness methods and flow-invariance for perturbed nonlinear semigroups. *An. Stiint. Univ. Iasi.* 27 (1981) 117-125.

[3] An existence result for a class of nonlinear evolution equations in Banach spaces. *Nonlinear Analysis* 6 (1982) 711-722.

Webb, G.F.

[1] Continuous nonlinear perturbations of linear accretive operators in Banach spaces. *J. Funct. Anal.* 10 (1972) 191-203.

Whittaker, E.T.

[1] *A Treatise on the Analytical Dynamics of Particles and Rigid Bodies.* Cambridge University Press (1961).

Yorke, J.A.

[1] Invariance for ordinary differential equations. *Math. Systems Theory* 1 (1967) 353-372.

[2] A continuous differential equation in Hilbert space without existence. *Funkcial. Ekvac.* 19 (1970) 19-21.

Zaidman, S.D.

[1] *Abstract Differential Equations.* (Pitman Research Notes in Mathematics, 36 (1979)).

Index